SERIES ON SEMICONDUCTOR
SCIENCE AND TECHNOLOGY

Series Editors

R. J. Nicholas   University of Oxford
H. Kamimura   University of Tokyo

Series on Semiconductor Science and Technology

1. M. Jaros: *Physics and applications of semiconductor microstructures*
2. V.N. Dobrovolsky and V. G. Litovchenko: *Surface electronic transport phenomena in semiconductors*
3. M.J. Kelly: *Low-dimensional semiconductors*
4. P.K. Basu: *Theory of optical processes in semiconductors*
5. N. Balkan: *Hot electrons in semiconductors*
6. B. Gil: *Group III nitride semiconductor compounds: physics and applications*
7. M. Sugawara: *Plasma etching*
8. M. Balkanski, R.F. Wallis: *Semiconductor physics and applications*
9. B. Gil: *Low-dimensional nitride semiconductors*
10. L. Challis: *Electron–phonon interactions in low-dimensional structures*
11. V. Ustinov, A. Zhukov, A. Egorov, N. Maleev: *Quantum dot lasers*
12. H. Spieler: *Semiconductor detector systems*
13. S. Maekawa: *Concepts in spin electronics*
14. S. D. Ganichev, W. Prettl: *Intense terahertz excitation of semiconductors*
15. N. Miura: *Physics of semiconductors in high magnetic fields*
16. A.V. Kavokin, J. J. Baumberg, G. Malpuech, F. P. Laussy: *Microcavities*
17. S. Maekawa, S. O. Valenzuela, E. Saitoh, T. Kimura: *Spin current*
18. B. Gil: *III-nitride semiconductors and their modern devices*
19. A. Toropov, T. Shubina: *Plasmonic Effects in Metal-Semiconductor Nanostructures*
20. B.K. Ridley: *Hybrid Phonons in Nanostructures*

Hybrid Phonons in Nanostructures

First Edition

B. K. Ridley

Professor Emeritus of Physics, University of Essex, UK

UNIVERSITY PRESS

OXFORD
UNIVERSITY PRESS

Great Clarendon Street, Oxford, OX2 6DP,
United Kingdom

Oxford University Press is a department of the University of Oxford.
It furthers the University's objective of excellence in research, scholarship,
and education by publishing worldwide. Oxford is a registered trade mark of
Oxford University Press in the UK and in certain other countries

© B.K. Ridley 2017

The moral rights of the author have been asserted

First Edition published in 2017

All rights reserved. No part of this publication may be reproduced, stored in
a retrieval system, or transmitted, in any form or by any means, without the
prior permission in writing of Oxford University Press, or as expressly permitted
by law, by licence or under terms agreed with the appropriate reprographics
rights organization. Enquiries concerning reproduction outside the scope of the
above should be sent to the Rights Department, Oxford University Press, at the
address above

You must not circulate this work in any other form
and you must impose this same condition on any acquirer

Published in the United States of America by Oxford University Press
198 Madison Avenue, New York, NY 10016, United States of America

British Library Cataloguing in Publication Data

Data available

Library of Congress Control Number: 2016946809

ISBN 978–0–19–878836–2

Preface

Crystalline nanostructures confine electrons and quantize their energies, which effect has led to the nomenclature quantum wells, quantum wires, and quantum dots. They also confine lattice waves. In the case of acoustic modes propagating in wave guides, this confinement is well understood by classical physics. This is not the case for optical modes of vibration. In acoustic wave guides and, indeed, in nanostructures, the normal modes of vibration are determined by the satisfaction of the classical connection rules—continuation of particle displacement and continuity of stress—at the boundaries. In order to satisfy these rules, a mode may have to combine with a mode of different polarization to form a hybrid. In many cases, a longitudinally polarized acoustic (LA) mode must liase with a transversely polarized acoustic (TA) mode to form a viable normal mode of the system. Such hybrid modes are commonplace in confining structures and have been known from the end of the nineteenth century. This is not the case for optical modes. Relative to acoustic modes, optical modes are but lately come, and their properties not as well defined or understood. This lacuna has been one of the motives for writing this book.

An understanding of optical modes in room-temperature devices is of some importance since they are, in polar structures, the principal source of electrical resistance. Acoustic modes are also important in that respect, but the shared frequencies between barrier and well make the confinement of acoustic modes at once more intricate and more open to simplification. On the one hand, reflection and transmission at the boundary lead to a rich family of mode patterns that include guided modes and interface waves. On the other hand, as regards the electron–phonon interaction, it may be sufficient for many purposes to disregard the intricacies entirely and treat the entire acoustic spectrum as bulk-like. This is not an easily justifiable option for optical modes, given the disparity of frequency between barrier and well that is commonly encountered. In that respect, optical modes present a problem. What exactly are the mechanical connection rules? In the case of polar modes, there are the usual electromagnetic boundary conditions as well as those mysterious rules associated with the elasticity of the lattice. Can the electromagnetic boundary conditions be sufficient? In other words, can the crystal be regarded simply as a dielectric continuum? For those interested solely in estimating the strength of the interaction between electrons and polar optical modes, the dielectric continuum (DC) model provides a simpler alternative to hybrid theory, an alternative that is not without some theoretical justification. Nevertheless, a crystal is not a simple dielectric continuum. If the physics of nanostructures is to see the semiconductor as a continuum, it must be a continuum that possesses both elastic and dielectric properties, inhabited by hybrid lattice vibrations, both acoustic and optical, along with confined electrons. These constitute the essential elements of the nanostructures and their interaction that will be described here.

Inevitably, such a description generates many equations, which many students of nanostructure physics may find somewhat indigestible. As one who prefers intuition to rigour (for better or worse), and who observes somewhat distantly the purely formal mathematical approach with some admiration, I have much sympathy with this attitude, but the student should know that the equations would be much more indigestible were they to portray a truly rigorous reality that took into account the natural anisotropy of semiconductor crystals. For simplicity, the hybrid modes that are described here are creatures of purely isotropic solids, in which modes are polarized purely longitudinally or purely transversely. Moreover, they are all long-wavelength modes, which allow a clear distinction to be made between optical and acoustic. Such approximations are acceptable for the Groups IV and III-V cubic semiconductors, but not for the hexagonal II-VI materials, which are highly anisotropic and, moreover, exhibit more than one optical mode. The book has been written with cubic semiconductors very much in mind.

Some parts of this book were written during and after moving house from Essex to Herefordshire (often to the despair of my wife). I suspect it has kept me sane during what most think of as one of the most traumatic events of life. Perhaps physics is to be recommended as a balm in troublesome times. My wife, bless her, doubts it.

Pembridge 2016

Contents

Acknowledgements	xi
Introduction	1
Prelude to Part 1	6

Part 1 Basics

1	Acoustic Modes	15
	1.1 Continuum Theory	15
	1.2 Equation of Motion	16
	1.3 Velocities	18
	1.4 Isotropic Case	19
	1.5 Inhomogenous Material	19
	1.6 Quantization	21
2	Optical Modes	24
	2.1 Introduction	24
	2.2 Microscopic Theory of the Diamond Lattice	25
	2.3 Decoupled Acoustic and Optical Equations	29
	2.4 Velocities	33
	2.5 Isotropy	34
	2.6 Inhomogeneous System	34
3	Polar Modes in Zinc Blende	37
	3.1 Polar Elements	37
	3.2 Polar Optical Modes	38
	3.3 Interface Modes	41
	3.4 Velocities	42
	3.5 Inhomogenous Material	43
	3.6 Piezoelectricity	43
4	Boundary Conditions	46
	4.1 Introduction	46
	4.2 Acoustic Modes	46
	4.3 Optical Modes	48
	4.4 Electromagnetic Boundary Conditions	54
5	Scalar and Vector Fields	55
	5.1 Introduction	55
	5.2 The Helmholtz Equation	55

	5.3 Cylinder	56
	5.4 Sphere	57

Part 2 Hybrid Modes in Nanostructures

6 Non-Polar Slab — 61

- 6.1 Boundary Conditions — 61
- 6.2 Acoustic Modes — 62
- 6.3 Optical Modes — 67

7 Single Heterostructure — 69

- 7.1 The Hybrid Model for Polar Optical Modes — 69
- 7.2 Remote Phonons — 72
- 7.3 Energy Normalization — 73
- 7.4 Reduced Boundary Condition — 74
- 7.5 Acoustic Hybrids — 75
- 7.6 Interface Acoustic Modes — 80

8 Quantum Well — 83

- 8.1 Triple Hybrid — 83
- 8.2 Energy Normalization — 88
- 8.3 Reduced Boundary Condition — 88
- 8.4 General Comments — 89
- 8.5 Barrier Modes — 90
- 8.6 Acoustic Modes — 91
- 8.7 Interface Acoustic Waves — 95
- 8.8 Guided Acoustic Waves — 96

9 Quantum Wire — 97

- 9.1 Introduction — 97
- 9.2 Cylindrical Coordinates — 98
- 9.3 Interface Modes — 101
- 9.4 Hybrid Modes in Polar Material — 103
- 9.5 Acoustic Stresses and Strains — 106
- 9.6 Free Surface — 108

10 Quantum Dot — 110

- 10.1 Introduction — 110
- 10.2 Spherical Coordinates — 110
- 10.3 Polar Double Hybrids — 114
- 10.4 Quantum Disc and Quantum Box — 115

Part 3 Electron–Phonon Interaction

11 The Interaction between Electrons and Polar Optical Phonons in Nanostructures: General Remarks — 119

    11.1 A Brief History — 119
    11.2 Dispersion — 121
    11.3 Coupled Modes and Hot Phonons — 122

12 Electrons — 124

    12.1 Confinement — 124
    12.2 Scattering Rate — 128

13 Scattering Rate in a Single Heterostructure — 129

    13.1 Scattering Rate — 129

14 Scattering Rate in a Quantum Well — 135

    14.1 Preliminary — 135
    14.2 Scattering Rate Associated with Quantum Well Modes — 135
    14.3 Scattering Rate Associated with Barrier Modes — 139
    14.4 General Remarks — 140

15 Scattering Rate in Quantum Wires — 142

    15.1 General Remarks — 142
    15.2 Scattering Rate — 142

16 The Electron–Phonon Interaction in a Quantum Dot — 145

    16.1 Preamble — 145
    16.2 Electron–Lattice Coupling — 145
    16.3 The Exciton — 148

17 Coupled Modes — 153

    17.1 Introduction — 153
    17.2 Long-Wavelength Modes — 153
    17.3 Beyond the Long-Wavelength Approximation — 156
    17.4 Screening in Quasi-2D Structures — 163
    17.5 Coupling to Hybrids — 170
    17.6 Quasi-1D Cylindrical Structures — 171
    17.7 Mobility — 172

18 Hot Phonon Lifetime — 173

    18.1 Introduction — 173
    18.2 Lifetime — 175
    18.3 Thermal Conductivity — 180

References — 185
Index — 189

Acknowledgements

The theory of hybrid optical modes presented here owes much to my colleagues at Essex and elsewhere. Mohamed Babiker was the first to direct my attention to mechanical as distinct from electrical boundary conditions. With him, I enjoyed a number of heretical discussions that focused on the choice of the scalar or the vector potential to describe interface modes. Nic Constantinou, Colin Bennett, and Nic Zakhlenuik made important contributions by illustrating the connection between hybrid and DC models in number of structures, and Martyn Chamberlain established the vital result that continuum hybrid models and computer-intensive lattice dynamical models produce the same dispersion relations. That tension between microscopic and continuum models, which was seen as a problem for both electron and lattice waves, was relaxed by Mike Burt in his envelope-function theory for electrons, and by Brad Foreman in his quasi-continuum theory for lattice waves. Those mechanical boundary conditions for optical modes have been a problem from the beginning, and the account given in this book is indebted to Brad Foreman's analytic theory of the relevant lattice dynamics. Without Angela Dyson's help the role of hot phonons and screening would not have been so clear. I am indebted for her many contributions during our collaboration in our research on electron transport in nanostructures, and to Paul Maki of the US Office of Naval Research for his support for this research.

I would also thank Sönke Adlung, Ania Wronski, Janet Walker, and Narmatha Vaithiyanathan for their support and editorial help.

Introduction

Advances in nanotechnology have produced structures a few molecular layers thick of crystalline semiconductors, and this has led to new challenges in the physics of solids. The semiconductors have been predominantly those of Group IV and Group III-V compounds, cubic with tetrahedral bonding. In particular, the confinement of electrons, resulting in the quantization of their motion, has produced novel electrical and optical properties and, as such, has defined the nomenclature as quantum wells, quantum wires, and quantum dots. The acoustic and optical waves of the lattice also suffer confinement, and the resultant change in their properties has to be taken into account in the treatment of the electron–phonon interaction. The electron–phonon interaction is central in determining both electrical and optical properties, so acquiring an understanding of the effect of confinement is of prime importance.

The quantization of the motion of the electron is well understood in terms of Schrödinger's equation and effective-mass theory (see e.g. Ridley 2009), and this topic will be touched upon in this book only briefly. The electron–phonon interaction itself is exhaustively treated elsewhere (e.g. Stroscio and Datta 2001), albeit solely in terms of a model of polar optical modes that regards the semiconductor simple as a dielectric continuum. The confinement of lattice waves cannot be described by such a model, and it is the purpose of this book to explore how confinement forces a hybridization of longitudinally and transversely polarized modes and how that affects the electron–phonon interaction.

Acoustic hybrids in slabs and wave-guides have been the subject of intense study for many years, particularly in connection with the exploitation of the piezoelectric effect (see e.g. Auld 1990). In spite of adopting the simplification of elastic isotropy and focusing on simple geometries, the solutions are very complicated, and especially so when piezoelectricity is fully taken into account. This has become a specialized field in its own right and beyond the scope of this book. We note briefly that the case of rectangular quantum wires has been described by Yu and colleagues (1994a, 1994b), and that for cylindrical wires by Beltzer (1988) and Stroscio (1989), and for quantum dots Sirenko and colleagues (1966). The study of the optical properties of quantum dots and exciton–phonon interactions has been recorded in a number of books (e.g. Datta 1995; Banyai and Koch 1993).

As regards the study of polar optical phonons, the focus has been generally on the interaction with electrons, the result of which has been to favour a simpler model than that necessitating hybrids. This, known as the dielectric continuum (DC) model, was used by Fuchs and Kliewer (1965) in their study of polar optical modes of a slab. The simplification comes from abandoning the necessity of satisfying mechanical boundary conditions and retaining only the electromagnetic boundary conditions. An argument for dispensing with mechanical boundary conditions is that dispersion is weak for long-wavelength modes and, as dispersion is a function of elastic properties, those boundary conditions associated with elastic properties, such as demanding the continuity of stress, cannot be important. The argument is invalid since it ignores the necessity for the restoring force between the two atoms in a unit cell to be continuous, which involves the variation of mass across the interface, as Akero and Ando (1989) were the first to point out. The DC model, strictly, is wrong, but its use in calculating the strength of the electron–phonon interaction more simply than by using the more accurate hybrid model finds practical justification in that the results obtained agree approximately with the results obtained from the hybrid model. An important theoretical justification has been given by Nash (1992). He showed that the scattering rate in the normal case of nearly degenerate LO modes is independent of the set of modes chosen, provided that the modes were complete and orthogonal, and at thermodynamic equilibrium. As a result, the study of hybrid optical modes in various nanostructures has been limited.

It is, nevertheless, the case that the modes chosen in the DC model are not the actual modes of the structure. The latter are revealed by Raman scattering (Sood et al. 1985) and are hybrids of LO, TO, and interface modes as required to satisfy both mechanical and electromagnetic boundary conditions (Ridley 1992, 1993; Trallero-Giner et al. 1992; Comas et al. 1993; Roca et al. 1994). Moreover, Raman scattering data (Sood et al. 1985; Mowbray et al. 1991) in GaAs show that the dispersion is essentially parabolic, in agreement with neutron scattering data (Strauch and Dorner 1990), which supports the result of continuum theory for moderate wave vectors. Nash has shown that the inclusion of dispersion does not invalidate the justification of the DC model, provided that the dispersion is weak. However, this cannot be the case when, as a result of the electron–phonon interaction or by optical excitation, the phonons become hot. How hot they get depends, of course, on frequency, and dispersion, which is affected by confinement, then becomes important. A study of hot hybrid phonons has yet to be carried out.

Solutions for hybrid optical phonons in various structures are as complicated as they are for acoustic modes. The case of polar optical hybrids in an AlAs/GaAs superlattice, studied by Chamberlain et al. (1993), is a good illustration. This work had the added usefulness in demonstrating that the results of analytic continuum theory showed excellent agreement with the computer-intensive numerical results of microscopic lattice dynamics models.

Whereas classical elasticity theory has been readily available for describing hybridized acoustic modes, this has not been the case for optical modes. A general, largely unanalysed assumption, has been to take the acoustic format as a model for the elasticity of optical modes, but it turns out that there are significant differences that can be seen to emerge naturally from the study of the microscopic dynamics of the tetrahedrally bonded lattice. The analytic model of lattice dynamics describing the individual motions of the two atoms in the unit cell, and their resolution into acoustic and optical vibrations, is not as well known as it should be, so it is given prior consideration here. It highlights the crucial differences, setting aside differences in frequency and polarity, between acoustic and optical waves, especially in inhomogeneous material, which, of course, all nanostructures are.

The book divides into three parts. The first part is concerned with the basic properties of acoustic and optical modes in the cubic tetrahedrally bonded lattice. The second is concerned with the application of hybrid theory to nanostructures, and the third with the interaction between electrons and polar optical phonons. A preliminary section introduces the usefulness of models of linear chains of atoms in exhibiting the salient properties of lattice waves.

In Part 1, Chapter 1 recalls the classical continuum theory of non-polar acoustic modes that defines elastic stress and strain, and which describes the natural anisotropy of the diamond lattice. It enters the quantum era with an account of energy normalization in terms of annihilation and creation operators. Chapter 2 gives an account of an analytic lattice dynamical model and its relation to envelope-function theory that serves as a foundation for discussing the properties of non-polar optical modes and their connection with the properties of acoustic modes. It also serves to quantify the anisotropy of propagating optical waves, and it establishes the conditions that the interatomic force constants must satisfy in order for the simplification of isotropy to be approximately valid. The important additional properties of polar material form the subject matter of Chapter 3, where it is seen that electric forces produce a fundamental distinction between longitudinal optical (LO) and transverse optical (TO) modes. Moreover, electromagnetic interface modes appear, and dielectric functions are defined that take into account the dispersion of each mode. Chapter 4 gives a detailed discussion of the boundary conditions that optical modes must satisfy. Unlike the case of acoustic modes, the connection rules for optical modes are heavily influenced by the various mass factors—reduced mass, difference in mass of the two atoms in the unit cell, etc.—and this makes the connection rules rather complicated. Chapter 5 ends Part 1 with a brief summary of vector field theory which is needed to describe modes in those systems that require the use of curvilinear coordinates. It gives an account of how the Helmholtz and Laplace equations that describe the motion of optical modes are to be handled in these cases that involve both scalar and vector fields.

Part 2 opens with Chapter 6, in which both acoustic and optical hybrid modes are seen to arise in a free-standing non-polar slab as a consequence of the boundary conditions—purely mechanical ones in this case. The free-standing slab provides a useful introduction to hybrid theory of both acoustic and optical phonons, especially as the boundary condition for optical modes simplifies to the requirement that the optical displacement vanish at the surface. One might expect a treatment of the polar slab to follow: instead, Chapter 7 focuses on optical mode hybrids associated with a single heterojunction. These consist of a linear combination of LO, TO, and interface modes that obey both the mechanical and electrical boundary conditions. Hybrid modes in a single heterojunction have received very little attention in the literature, so Chapter 7 contains a description of both acoustic and optical modes. The quantum well was the first nanostructure to be analysed in terms of hybrid polar optical modes, and an account is given in Chapter 8 as well as an account of the various families of acoustic modes, interface and guided. This concludes our account of hybrids in structures that can be described in terms of Cartesian coordinates.

In our treatment of cylindrical quantum wires and spherical quantum dots it is natural to use appropriate curvilinear coordinates and to describe phonon modes in terms of scalar and vector potentials. Chapter 9 describes hybrids in a cylindrical wire and Chapter 10 describes hybrids in a spherical dot.

Where polar effects produce a significant difference in LO and TO frequencies (which is the usual case) it is possible to modify things to produce an effective hybrid theory of polar optical modes that is much simpler to apply. This involves the assumption that the TO component is effectively confined to the interface and has no effect other than eliminating the in-plane or tangential displacement. The mechanical boundary condition therefore reduces to the elimination of the displacement normal to the interface. This simplified model is used to describe the interaction between electrons and polar optical phonons. Part 3 opens with Chapter 11, which contains a brief history of the electron–phonon interaction in nanostructures, a reminder of the effects of lattice dispersion, and an overview of possible simple models that takes into account hot phonons and coupled modes. Chapter 12 describes the confinement of electrons which, compared with the confinement of phonons, has been blessedly free of controversy. The following Chapters 13, 14, and 15 contain accounts of the lifetime of an electron in a particular state as determined by the scattering rate in the systems: the single heterostructure, the quantum well, and the quantum wire.

The principal technological importance of the quantum dot is grounded in its optical, rather than its electrical, properties. An electron that appears in the quantum dot by photexcitation is accompanied by a hole in the valence band, to which it is attracted by the Coulomb force, resulting in the formation of some sort of exciton. The interaction with phonons is therefore more complex in nanostructures as a consequence of the simultaneous actions of coulomb and spatial confinement. Moreover, the absorption and emission spectra, unlike the case in bulk material, are affected by the anharmonic interaction with phonons that results from confinement, all of which makes the study of quantum dots a study in its own right. In Chapter 16 our discussion of the electron–phonon interaction is limited to the highlighting of the main features.

By their nature, modern high-power devices have high densities of electrons, and this introduces new effects. Plasma waves exist that can couple with the polar LO modes, which has the effect of changing the dispersion relation between frequency and wave vector and, further, screening the interaction with single electrons. In Chapter 17 we give a brief account of the properties of coupled plasmon–phonon modes. Another phenomenon associated with high-power devices is the production of hot phonons, which occurs when the rate of emission of optical phonons by the electrons far exceeds the rate of annihilation.

Phonons decay into other phonons via the anharmonic interaction, and it is this interaction that determines the lifetime of a phonon. In Chapter 18, we describe the mechanics of optical phonon production and decay. The anharmonic interaction is also active in determining the thermal conductivity, this time associated with acoustic modes. We address the question of the effect of hybridization and confinement in nanostructures on the thermal conductivity.

Prelude to Part 1

Space is three dimensional, and so is the matter that resides in it. Matter is made up of atoms, so it has a point-like structure, but it can also behave is if it were a continuum. It therefore has microscopic and macroscopic properties. As a consequence, the physical properties of matter are written in terms of scalar, vector, and tensor quantities that obey differential equations, involving many of the transcendental functions of mathematics, all of which can reach a high degree of complexity, even in the well-ordered structure of crystals. Or it can be described in terms of that well-ordered structure, taking into account the dynamic properties of atoms. The description of the vibrational modes of crystals is no exception, especially when those crystals have nanometre dimensions. Fortunately, it is possible to discover some vital properties of three-dimensional crystals quite simply by considering the vibrational properties of a linear chain of atoms, and this provides a useful prelude to Part 1.

We consider first a line of identical, equally spaced, atoms (Fig. 1a). The atoms are labelled by their position in the chain—$n-1, n, n+1$, etc. We suppose the force that acts on the nth atom is simply proportional to the displacements, u, from their equilibrium positions of the atom and its nearest neighbours:

$$F = \alpha(u_{n+1} - u_n) - \alpha(u_n - u_{n-1}) \qquad (1)$$

where α is the force constant. The equation of motion of the nth atom is

$$M\ddot{u}_n = \alpha(u_{n+1} + u_{n-1} - 2u_n) \qquad (2)$$

where M is the mass of an atom. This equation has a wave-like solution:

$$u_n = Ae^{i(kna - \omega t)} \qquad (3)$$

where A is an amplitude determined by a process of energy normalization that we can ignore for now. Substitution into equation (2) gives a dispersion relation, that is, a relation between angular frequency and wave vector:

$$-M\omega^2 = \alpha(e^{ika} + e^{-ika} - 2)$$
$$\omega^2 = \frac{2\alpha}{M}(1 - \cos ka) = \frac{4\alpha}{M}\sin^2(ka/2)$$
$$\omega = \left(\frac{4\alpha}{M}\right)^{1/2} \sin(|k|a/2) \qquad (4)$$

Figure 1 *Linear chain of atoms: (a) Monatomic model, all have the same mass. (b) Diatomic model, even number atoms have mass M_A, odd number atoms have mass M_B.*

Real values of the frequency are given by $0 \leq |k| \leq \pi/a$; beyond the maximum, the wave vector merely duplicates the frequency.

The question arises: what are the conditions at the extremities of the chain? This question raises the issue of boundary conditions, a vital issue in the description of waves in nanostructures. Naturally, these conditions will depend on the properties of the matter lying beyond the extremities of the chain, and there can be no simple answer. The problem can be ducked by imagining the chain to fold itself into an arc of a circle, so that the atom $n = 0$ is identical to atom $n = N$, where N is the total number of atoms in the chain. The finite length of the chain is now $L = Na$, and therefore the smallest value that the wave vector can have is $k_{min} = 2\pi/L = 2\pi/Na$. Values of the wave vector are limited to the region $\frac{2\pi}{Na} \leq |k| \leq \frac{\pi}{a}$, which is one-dimensional equivalent of the Brillouin zone in three-dimensional crystals. The total number of modes is N, so the magnitude of the wave vector is limited to values equal to $\pm 2\pi p/Na$, where p is an integer and $p_{max} = N/2$.

It is sometimes convenient to ignore the atomic structure and think in terms of travelling waves in a continuum, $0 \leq x \leq L$, with

$$u(x) = Ae^{i(kx-\omega t)} \tag{5}$$

Nevertheless, the atomic nature of the lattice demands that the number of modes equals the number of atoms. Integration now replacing sum, the number of modes is given by

$$N = \int_{-\pi/a}^{\pi/a} dk (L/2\pi) \tag{6}$$

Most of the semiconductor crystals of interest have a lattice with a basis, meaning that more than one atom inhabits each unit cell. A simple one-dimensional model of a lattice with two different sorts of atom is shown in Figure 1b. A atoms, each with mass M_A, reside on even number sites, B atoms, each with mass M_B, on odd number sites. The equations of motion are

$$M_A \ddot{u}_{2n} = \alpha(u_{2n+1} + u_{2n-1} - 2u_{2n})$$
$$M_B \ddot{u}_{2n+1} = \alpha(u_{2n+2} + u_{2n} - u_{2n+1}) \tag{7}$$

with

$$u_{2n} = Ae^{i(k2na-\omega t)}, \quad u_{2n+1} = Be^{i(k(2n+1)-\omega t)} \tag{8}$$

Assuming a common frequency results in the equations:

$$A(M_A\omega^2 - 2\alpha) + B2\alpha \cos ka = 0$$
$$A2\alpha \cos ka + B(M_B\omega^2 - 2\alpha) = 0 \qquad (9)$$

Solutions exist provided that the determinant of the coefficients of A and B vanishes. This requirement leads to the dispersion relation

$$\omega^2 = \alpha \left[\frac{1}{\overline{M}} \pm \left(\frac{1}{\overline{M}^2} - \frac{4\sin^2 ka}{M_A M_B} \right)^{1/2} \right] \qquad (10)$$

Here $\overline{M} = M_A M_B/(M_A + M_B)$ is the reduced mass. The long-wavelength solutions are

$$\omega_+^2 = 2\alpha \left[\frac{1}{\overline{M}} - \frac{(ka)^2}{M} \right], \quad A_+ = -\frac{M_B}{M_A} B_+ \left[1 - \frac{M_A - M_B}{2M}(ka)^2 \right]$$
$$\omega_-^2 = 2\alpha \frac{(ka)^2}{M}, \quad A_- = B_- \left[1 + \frac{M_A - M_B}{2M}(ka)^2 \right] \qquad (11)$$

where $M = M_A + M_B$. The upper frequency corresponds to optical modes, the lower to acoustic modes. In the long-wavelength limit, $(ka)^2 \ll 1$, the relations between the amplitudes are

$$A_+ = -\frac{M_B}{M_A} B_+, \quad A_- = B_- \qquad (12)$$

The energies are

$$E_{opt} = \omega^2 (M_A A_+^2 + M_B B_+^2) = \omega^2 \overline{M} u^2$$
$$E_{ac} = \omega^2 (M_A A_-^2 + M_B B_-^2) = \omega^2 M U^2 \qquad (13)$$

The acoustic displacement is that of the centre of gravity:

$$U = \frac{M_A}{M} A_- + \frac{M_B}{M} B_- \qquad (14)$$

The optical displacement is the relative displacement of the two atoms:

$$u = A_+ - B_+ \qquad (15)$$

(as the substitution of equation (12) into equation (13) shows).

It is worth remarking that the one-dimensional model can only depict longitudinally polarized modes; transversely polarized modes, by definition, do not exist.

The connection rules at an interface were well established by classical physics, but only for acoustic waves. No elastic continuum theory of optical modes existed until the last decade of the twentieth century, and that lacuna became a hindrance to a satisfactory understanding of the electron–phonon interaction in modern nanostructures. In particularly, it was not clear what boundary conditions optical modes had to obey. For want of anything better, it was often assumed that the acoustic connection rules would suffice, but Akero and Ando (1989) analysed a simple linear chain model and showed that this assumption was incorrect, and that it was necessary to take into account the difference of ionic masses. Mass does not enter the acoustic conditions, but it became clear from the analysis of the microscopic lattice dynamics that mass was, indeed, a factor.

In a linear chain, A, consisting of atoms with two different masses, is joined to a similar chain, B, with atoms having masses that are different. We suppose the interface lies between atoms n and $n+1$. For simplicity, we will assume that the force constants are the same in A and B, but for the moment we will distinguish the force constant acting across the interface. (There is no problem in being more sophisticated here, but unnecessary.) We consider the equations of motion of the atom n with displacement u_n and of atom $n+1$ with displacement u_{n+1}:

$$-M_{A1}\omega^2 u_n^2 = \alpha^*(u_{n+1} - u_n) + \alpha(u_{n-1} - u_n)$$
$$-M_{B1}\omega^2 u_{n+1}^2 = \alpha(u_{n+2} - u_{n+1}) + \alpha^*(u_n - u_{n+1}) \qquad (16)$$

Away from the interface these equations would be

$$-M_{A1}\omega^2 u_n^2 = \alpha(u_{n+1} - u_n) + \alpha(u_{n-1} - u_n)$$
$$-M_{B1}\omega^2 u_{n+1}^2 = \alpha(u_{n+2} - u_{n+1}) + \alpha(u_n - u_{n+1}) \qquad (17)$$

With $\alpha^* = \alpha$ the connection rules become, simply,

$$u_{n+1}^A = u_{n+1}^B$$
$$u_n^A = u_n^B \qquad (18)$$

We first consider an acoustic mode. Let the amplitude of the mode in crystal A be U_A and that in crystal B, U_B. We imagine that the phase of the wave is continuous in space, so that we may take

$$U(x) = U(x_0) + (x - x_0)\frac{\partial U}{\partial x} \qquad (19)$$

Let us take the interface half way between u_n and u_{n+1}, and define $\nabla U = a\frac{\partial U}{\partial x}$, where a is the interatomic distance. Let U_A be the amplitude of the wave in A at the interface, and U_B the amplitude of the wave in B. Then, according to the connection rules of equation (18), we have

$$u^A_{n+1} = U_A + (1/2)\nabla_A U = u^B_{n+1} = U_B + (1/2)\nabla_B U$$
$$u^A_n = U_A - (1/2)\nabla_A U = u^B_n = U_B - (1/2)\nabla_B U \tag{20}$$

In terms of a transfer matrix, the connection rules become

$$\begin{vmatrix} U_A \\ \nabla_A U \end{vmatrix} = T_{AB} \begin{vmatrix} U_B \\ \nabla_{AB} U \end{vmatrix}$$

$$T_{AB} = \begin{vmatrix} 1 & 0 \\ 0 & 1 \end{vmatrix} \tag{21}$$

This is exactly the classical result (given the mass approximation).

The connection rules for the optical modes are obtained in the same way, but taking into account the relation between the amplitudes as given in equation (12) (which is where the masses enter):

$$u^A_{n+1} = \frac{M^A_{n+1}}{M_A}u_A + \frac{M^A_{n+1}}{M_A}(1/2)\nabla_A u = u^B_{n+1} = \frac{M^B_{n+1}}{M_B}u_B + \frac{M^B_{n+1}}{M_B}(1/2)\nabla_B u$$
$$u^A_n = \frac{M^A_n}{M_A}u_A - \frac{M^A_n}{M_A}(1/2)\nabla_A u = u^B_n = \frac{M^B_n}{M_B}u_B - \frac{M^B_n}{M_B}(1/2)\nabla_B u \tag{22}$$

$$\begin{vmatrix} u_A \\ \nabla_A u \end{vmatrix} = T_{AB} \begin{vmatrix} u_B \\ \nabla_{AB} u \end{vmatrix}$$

$$T_{AB} = \begin{vmatrix} \frac{1}{2}(R_1 + R_2) & \frac{1}{4}(R_2 - R_1) \\ (R_2 - R_1) & \frac{1}{2}(R_1 + R_2) \end{vmatrix} \tag{23}$$

$$R_1 = \frac{M^B_{n+1}}{M^A_{n+1}}\frac{M_A}{M_B} \quad R_2 = \frac{M^B_n}{M^A_n}\frac{M_A}{M_B} \tag{24}$$

Unlike the acoustic case, neither n nor u nor its gradient is continuous in general. They are continuous, however, in the case that the masses of the two atoms in the unit cell are equal, as they are in Group IV semiconductors, for then $R_1 = R_2$.

The atoms in the chains that we have been considering are non-polar, and insofar as that is the case, the foregoing analyses refer to Group IV materials. The replacement of neutral atoms by ions in the chain introduces a new factor into the description of the force constants. The three-dimensional branches of TO and

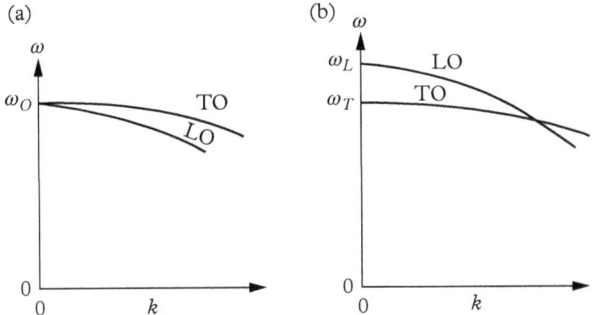

Figure 2 *Dispersion of (a) non-polar and (b) polar optical modes.*

LO modes are shown in Figure 2. It is no longer possible to assume the mass approximation. The force between the ions consists of two parts—a non-polar component and an electric component. The non-polar component in the mass approximation is, for similar lattices, effectively a universal constant, but the polar component depends on the ionic charge which is a property of the ion and whose magnitude is different for different ions. Moreover, connection rules must now include the classical electric connection rules between different dielectric media.

Another aspect of polarity is the piezoelectric effect associated with acoustic modes. With ions fixed firmly in their lattice positions the overall polarity is zero, but an applied mechanical stress may shift the ions from their equilibrium positions and produce an internal electric field. Equally well, an applied electric field, acting on the ions, can disturb the ions and thereby introduce internal mechanical strains. Piezoelectric effects of these sorts can occur only in lattices that lack a centre of symmetry, but this is true of all the semiconductors of interest, which are all anisotropic to some degree. The technological interest is focused on travelling acoustic wave devices—surface acoustic waves, acoustic wave guides, etc. In the case of room-temperature nanostructure devices, the piezoelectric interaction between acoustic phonons and electrons is weak and usually neglected. Neglecting the piezoelectric effect as regards the interaction with electrons allows us to feel a sense of self-consistency by assuming that all the nanostructures with which we will deal are blessedly isotropic and therefore non-piezoelectric.

Hexagonal semiconductors present more of a problem. Not only the properties along a direction in the basal plane of the lattice are expected to be different from those along the *c*-axis, but crystal-like GaN has four ions in the unit cell, and therefore more modes of lattice vibration. A linear chain with four ions per unit cell will produce four equations of motion with frequencies that are the solutions of a quartic equation: three sorts of optical mode plus an acoustic mode. An account of hybrid modes in nanostructures of hexagonal material is beyond the scope of this book.

Part 1

Basics

1
Acoustic Modes

1.1 Continuum Theory

The classical account of acoustic modes in a non-polar crystal regards the crystal lattice as a continuum whose infinitesimal elements undergo distortion in response to the action of internal stresses. An elementary parallelepiped of the material experiences a stress (force per unit area) on each of its faces, which can be of two sorts. If the stress is normal to the surface it produces compression or tension, depending on sign; if the stress is along the surface it produces shear. The resulting distortion is quantified in terms of elastic strain, defined as the displacement per unit length. Assuming that the stress is not too large, we can relate the strain linearly to the stress via an elastic constant: Hooke's law. Since a stress acts in a particular direction and on a particular surface, it is a second-order tensor, $T_{\alpha\beta}$. The strain is also a second-order tensor, $S_{\lambda\mu}$, since it is a displacement, u, in one direction relative to another direction. This makes the elastic coefficient a fourth-order tensor, $c_{\alpha\beta\lambda\mu}$. Stresses do not change the amount of matter, so a compressional stress will reduce the dimension in the normal direction and at the same time expand the dimensions in the orthogonal directions and even produce shear strains. Thus, in general, all possible strains are involved. Hooke's law can be stated in the form

$$T_{\alpha\beta} = \sum_{\lambda\mu} c_{\alpha\beta\lambda\mu} S_{\lambda\mu} \tag{1.1}$$

with

$$S_{\lambda\mu} = \partial u_\lambda / \partial \mu \tag{1.2}$$

For example, a force acting in the x direction on the x surface is described by the stress

$$\begin{aligned}T_{xx} = {}& c_{xxxx}S_{xx} + c_{xxyy}S_{yy} + c_{xxzz}S_{zz} + c_{xxxy}S_{xy} + c_{xxxz}S_{xz} \\ & + c_{xxyx}S_{yx} + c_{xxzx}S_{zx} + c_{xxyz}S_{yz} + c_{xxzy}S_{zy}\end{aligned} \tag{1.3}$$

Hybrid Phonons in Nanostructures. First Edition. B.K. Ridley. © B.K. Ridley 2017.
Published in 2017 by Oxford University Press. DOI: 10.1093/acprof:oso/9780198788362.001.0001

A force acting in the x direction on the y surface is described by the stress

$$T_{xx} = c_{xyxx}S_{xx} + c_{xyyy}S_{yy} + c_{xyzz}S_{zz} + c_{xyxy}S_{xy} + c_{xyxz}S_{xz}$$
$$+ c_{xyyx}S_{yx} + c_{xxzx}S_{zx} + c_{xyyz}S_{yz} + c_{xyzy}S_{zy} \qquad (1.4)$$

Corresponding expressions exist for $T_{yy}, T_{zz}, T_{xz}, T_{yx}, T_{zx}, T_{yz}, T_{zy}$. Group IV crystals (diamond, silicon, germanium), Group III-V (AlAs, GaAs, etc.), and some Group II-VI (e.g. ZnS) have cubic lattices whose symmetry reduces the number of distinct, non-zero, elastic constants to three:

$$c_{xxxx} = c_{yyyy} = c_{zzzz}$$
$$c_{xxyy} = c_{xxzz} = c_{yyxx} = c_{zzxx} = c_{yyzz} = c_{zzyy}$$
$$c_{xyxy} = c_{xyyx} = c_{yzyz} = c_{yzzy} + c_{zxzx} + c_{zxxz} \qquad (1.5)$$

The rest are zero. The III-V and II-VI compounds are polar, which introduces electric fields proportional to displacement and strain. This will be discussed in Chapter 3. Here we will continue to focus on the non-polar elements of acoustic modes, these being mechanically the most important.

It is convenient to simplify the suffixes using reduced notation, in which $xx = 1, yy = 2, zz = 3, yz = 4, zx = 5, xy = 6$. The three non-zero elastic constants are then c_{11}, c_{12}, c_{44}. The relation between stress and strain can then be expressed:

$$T_1 = c_{11}S_1 + c_{12}(S_2 + S_3)$$
$$T_2 = c_{11}S_2 + c_{12}(S_3 + S_2)$$
$$T_3 = c_{11}S_3 + c_{12}(S_1 + S_2)$$
$$T_4 = c_{44}S_4$$
$$T_5 = c_{44}S_5$$
$$T_6 = c_{44}S_6 \qquad (1.6)$$

The strains are

$$S_i = \frac{\partial u_i}{\partial x_i}, \quad i = 1, 2, 3$$
$$S_4 = \frac{\partial u_2}{\partial x_3} + \frac{\partial u_3}{\partial x_2} = S_{23} + S_{32}$$
$$S_5 = S_{31} + S_{13}$$
$$S_6 = S_{12} + S_{21} \qquad (1.7)$$

1.2 Equation of Motion

We first remind ourselves that it is the difference of stress, not stress itself, that is responsible for providing a net force. If we imagine a cube experiencing a stress

T applied to one side then there will be an equal and opposite stress applied to the opposite side to prevent a mass movement of the cube. In this case the force per unit area, which is the stress, applied to one face is balanced by that on the other. Only if T is a function of distance will there be a net force, F. Thus if the stress is T at x and $T + (dT/dx)dx$ at $x + dx$, the net force is

$$F = -\frac{\partial T}{\partial x} dx dy dz \qquad (1.8)$$

Taking account of the fact that shear stresses produce displacements at right angles, we can write the equation of motion for acoustic displacements in one direction as follows:

$$\rho \ddot{U} = \frac{\partial T_1}{\partial x_1} + \frac{\partial T_6}{\partial x_2} + \frac{\partial T_5}{\partial x_3} \qquad (1.9)$$

where ρ is the mass density and the right-hand side is the restoring force. We want to use u for the displacement for optical modes and U for the displacement for acoustic modes. Substituting strain, we get

$$\rho \ddot{U}_1 = c_{11}\frac{\partial S_1}{\partial x_1} + c_{12}\frac{\partial (S_2 + S_3)}{\partial x_1} + c_{44}\left(\frac{\partial S_6}{\partial x_2} + \frac{\partial S_5}{\partial x_3}\right) \qquad (1.10)$$

Replacing the strains by displacements and rearranging leads to

$$\rho \ddot{U}_1 = c_{44}\left(\frac{\partial^2 U_1}{\partial x_1^2} + \frac{\partial^2 U_1}{\partial x_2^2} + \frac{\partial^2 U_1}{\partial x_3^2}\right) + (c_{12} + c_{44})\frac{\partial}{\partial x_1}\left(\frac{\partial U_1}{\partial x_1} + \frac{\partial U_2}{\partial x_2} + \frac{\partial U_3}{\partial x_3}\right)$$

$$+ (c_{11} - c_{12} - 2c_{44})\frac{\partial^2 U_1}{\partial x_1^2} \qquad (1.11)$$

This can be generalized:

$$\rho \ddot{\mathbf{U}} = c_{44}\nabla^2 \mathbf{U} + (c_{12}+c_{44})\nabla(\nabla \cdot \mathbf{U}) + (c_{11}-c_{12}-2c_{44})\left(\hat{\mathbf{x}}\frac{\partial^2 U_1}{\partial x_1^2} + \hat{\mathbf{y}}\frac{\partial^2 U_2}{\partial x_2^2} + \hat{\mathbf{z}}\frac{\partial^2 U_3}{\partial x_3^2}\right)$$

$$(1.12)$$

where $\hat{\mathbf{x}}, \hat{\mathbf{y}},$ and $\hat{\mathbf{z}}$ are unit vectors along the axes. We note that the motion is different for different directions, evidence of elastic anisotropy. Also, note that pure longitudinal and transverse polarizations are limited to modes propagating along the principal axes: for general direction the polarizations are mixed.

1.3 Velocities

For a travelling wave of the form

$$\mathbf{U} = \mathbf{a} e^{i(\mathbf{k}\cdot\mathbf{r}-\omega t)} \tag{1.13}$$

equation (1.12) becomes

$$\rho\omega^2 \mathbf{a} = c_{44} k^2 \mathbf{a} + (c_{12} + c_{44})\mathbf{k}(\mathbf{k}\cdot\mathbf{a}) + c_0(\hat{\mathbf{x}} k_1^2 a_1 + \hat{\mathbf{y}} k_2^2 a_2 + \hat{\mathbf{z}} k_3^2 a_3) \tag{1.14}$$

$$c_0 = c_{11} - c_{12} - 2c_{44} \tag{1.15}$$

Define a velocity v by $\omega = vk$ and express the components of $\mathbf{k}$ in terms of direction cosines α, β, γ. This allows us to eliminate k^2 and resolve equation (1.14) into three simultaneous equations for the components of $\mathbf{a}$, the polarization vector:

$$a_1[c_{44} + \alpha^2(c_{12} + c_{44} + c_0) - \rho v^2] + a_2(c_{12} + c_{44})\alpha\beta + a_3(c_{12} + c_{44})\alpha\gamma = 0$$
$$a_1(c_{12} + c_{44})\alpha\beta + a_2[c_{44} + \beta^2(c_{12} + c_{44} + c_0) - \rho v^2] + a_3(c_{12} + c_{44})\beta\gamma = 0$$
$$a_1(c_{12} + c_{44})\alpha\gamma + a_2(c_{12} + c_{44})\beta\gamma + a_3[c_{44} + \gamma^2(c_{12} + c_{44} + c_0) - \rho v^2] = 0 \tag{1.16}$$

Solutions exist provided that the determinant of the coefficients of the a_i vanishes.

As examples we consider propagation along each of the principal axes. The [100] direction implies that $\alpha = 1, \beta = \gamma = 0$. The determinant is

$$\begin{vmatrix} a_1[(c_{12} + 2c_{44} + c_0) - \rho v^2] & & \\ & a_2[c_{44} - \rho v^2] & \\ & & a_3[c_{44} - \rho v^2] \end{vmatrix} \tag{1.17}$$

There are three solutions, corresponding to one purely longitudinally polarized acoustic (LA) wave plus two purely transversely polarized acoustic (TA) waves. The velocities of the LA and TA waves are obtained from

$$\rho v_L^2 = (c_{12} + 2c_{44} + c_0) = c_{11}$$
$$\rho v_T^2 = c_{44}$$
$$\rho v_T^2 = c_{44} \tag{1.18}$$

In the [110] direction ($\alpha = \beta = 1/\sqrt{2}, \gamma = 0$) the velocities are found from

$$\rho v_L^2 = (c_{12} + 2c_{44} + c_0/2)$$
$$\rho v_T^2 = c_{44} + c_0/2$$
$$\rho v_T^2 = c_{44} \tag{1.19}$$

In the [111] direction ($\alpha = \beta = \gamma = 1/\sqrt{3}$) the results are

$$\rho v_L^2 = (c_{12} + 2c_{44} + c_0/3)$$
$$\rho v_T^2 = c_{44} + c_0/3$$
$$\rho v_T^2 = c_{44} + c_0/3 \tag{1.20}$$

The elastic energy density associated with strain is the classical expression:

$$V = \frac{1}{2}c_{11}(S_1^2 + S_2^2 + S_3^2) + c_{12}(S_2 S_3 + S_3 S_1 + S_1 S_2) + \frac{1}{2}(S_4^2 + S_5^2 + S_6^2) \tag{1.21}$$

1.4 Isotropic Case

It is often very convenient to make the simplifying assumption that the crystal is elastically isotropic, which implies that

$$c_{11} - c_{12} - 2c_{44} = 0 \tag{1.22}$$

By eliminating c_{12}, the energy density becomes

$$V = \frac{1}{2}c_{11}(\nabla \cdot \mathbf{U})^2 + \frac{1}{2}c_{44}(\nabla \times \mathbf{U})^2 \tag{1.23}$$

Since $\nabla \cdot \mathbf{U} = 0$ for transversely polarized modes, the first term describes the energy of longitudinally polarized modes. Similarly, $\nabla \times \mathbf{U} = 0$ for longitudinally polarized modes, so the second term is the energy of transversely polarized modes. The energy then is simply the sum of the energies of the longitudinal and transverse modes. The assumption of elastic isotropy decouples the longitudinal and transverse modes, providing a simplification, which we will find useful.

1.5 Inhomogenous Material

In crystalline nanostructures there are interfaces where the crystal properties change more or less abruptly. The reflection and transmission of waves through an interface must be determined by the conditions that must hold at the interface. These conditions are the continuity of amplitude and stress along with the classic electromagnetic boundary conditions when electric fields are involved—the continuity of the tangential component of field and the continuity of the electric displacement along the normal to the interface—and the frequency must be the same on both sides if there is transmission.

Acoustic Modes

The mechanical boundary condition involving stress means that the change in the atomic force constants across the interface must be taken into account. The general expression for the equation of motion in the vicinity of the interface becomes

$$\omega^2 \rho U_\alpha = -(c_{\alpha\lambda\beta\mu} U_{\beta,\mu})_{,\lambda} \tag{1.24}$$

In this equation we use the space-saving convention of the comma defining differentiation, thus $U_{\beta,\mu} = \frac{\partial U_\beta}{\partial x_\mu}$

Two properties conspire to allow the approximation in which a change in elastic constant across an interface can be regarded as insignificant: one is the observation that the magnitude of an elastic constant is roughly proportional to the lattice constant (e.g. Adachi 1985), and the second is the practical desirability of crystal growth for adjacent materials be lattice matched. To take an example—AlAs/GaAs—we compare magnitudes in Table 1.1.

Without too much error we can take the acoustic equation to be

$$\omega^2 \rho U_\alpha = - c_{\alpha\lambda\beta\mu} U_{\beta,\mu\lambda} \tag{1.25}$$

regarding the elastic constants to be the same in each of the lattice-matched components of the interface. Writing this out explicitly for the x component, we obtain

$$\omega^2 \rho U_1 = - c_{11} U_{x,xx} - c_{12}(U_{y,y} + U_{z,z}) - c_{44}(U_{x,yy} + U_{y,xy} + U_{x,zz} + U_{z,xz}) \tag{1.26}$$

But, even with this simplification, the effect of the discontinuity of mass density and the resultant discontinuity of velocity at a boundary between two semiconductors introduce significant changes to the properties of acoustic waves. This is because the classical elastic boundary conditions, entailing the continuity of amplitude and stress, cannot, in general, be satisfied by the familiar transverse acoustic (TA) and longitudinal acoustic (LA) modes of bulk material. As a result, an interface has the effect of forcing a degree of hybridization of polarization, producing mixed TA and LA waves of the sort that were completely decoupled in isotropic bulk material.

Table 1.1 *Elastic constants (10^{10} Nm^{-2}).*

	AlAs	GaAs
c_{11}	12.0	11.9
c_{12}	5.7	5.4
c_{44}	5.9	6.0

The new physics that this introduces in a range of nanostructures has been explored in depth by a number of authors. Acoustic modes in a free-standing slab have been exhaustively studied by Bannov and colleagues (1995) who have provided numerical solutions to the dispersion and have illustrated the possibility of modes becoming localized at the surfaces. Quantum-well modes have been described by Wendler and Grigoryan (1988), again finding it necessary to establish the dispersion of the modes numerically. The study of acoustic waves in rods of rectangular cross section goes back to the work of Morse (1948, 1950). More recent accounts of modes in rectangular quantum wires have been given by Yu and colleagues (1994a) and in cylindrical quantum wires by Yu and colleagues (1994b, 1996) and Stroscio and colleagues (1996). The situation in quantum dots, rectangular faced and spherical, was studied by Stroscio and colleagues (1994). An excellent account of acoustic waves in wave guides can be found in the book by Auld (1990), and for some nanostructures in the book by Stroscio and Datta (2001).

In view of the extensive literature that exists in this field, my focus in this book will be limited to a brief description of modes in a free-standing slab (Chapter 6), in a single heterojunction (Chapter 7), and in a quantum wire (Chapter 8), sufficient to illuminate the main features of acoustic mode confinement.

The main focus in later chapters, however, will be on optical polar modes and their interaction with electrons.

1.6 Quantization

The small-amplitude vibrations of the lattice are those of the simple harmonic oscillator, whose quantization is well known. The Hamiltonian is of the form

$$\hat{H} = \frac{\hat{p}^2}{2m} + \frac{1}{2}m\omega^2 \hat{q}^2, \quad [\hat{q}, \hat{p}] = i\hbar \quad (1.27)$$

The Hamiltonian, the momentum, and spatial coordinates are operators $\hat{H}$, $\hat{p}$, and $\hat{q}$ respectively. The energy of an oscillator state is quantized in units of $\hbar\omega$, and is characterized by the number of such quanta and it is therefore convenient to describe a state by the ket $|n\rangle$, where n is the number of quanta, especially where interactions that change that number are considered. We define annihilation operator a and creation operator $a^\dagger$ in terms of the momentum and displacement operators as follows:

$$\hat{a} = \left(\frac{1}{2m\hbar\omega}\right)^{1/2} (m\omega\hat{q} + i\hat{p})$$

$$\hat{a}^\dagger = \left(\frac{1}{2m\hbar\omega}\right)^{1/2} (m\omega\hat{q} - i\hat{p}) \quad (1.28)$$

Conversely,

$$\hat{q} = \left(\frac{\hbar}{2m\omega}\right)^{1/2} (a^\dagger + a)$$

$$\hat{p} = i\left(\frac{m\hbar\omega}{2}\right)^{1/2} (a^\dagger - a) \quad (1.29)$$

It is then true that

$$\hat{a}\hat{a}^\dagger = \frac{1}{\hbar\omega}\left(\hat{H} + \frac{1}{2}\hbar\omega\right)$$

$$\hat{a}^\dagger\hat{a} = \frac{1}{\hbar\omega}\left(\hat{H} - \frac{1}{2}\hbar\omega\right) \quad (1.30)$$

Taking the sum gives

$$\hat{H} = \frac{\hbar\omega}{2}(\hat{a}\hat{a}^\dagger + \hat{a}^\dagger\hat{a}) \quad (1.31)$$

Taking the difference gives

$$\hat{a}\hat{a}^\dagger - \hat{a}\hat{a}^\dagger = [\hat{a}, \hat{a}^\dagger] = 1 \quad (1.32)$$

Equation (1.31) becomes

$$\hat{H} = \hbar\omega\left(\hat{a}^\dagger\hat{a} + \frac{1}{2}\right) \quad (1.33)$$

The product $\hat{a}^\dagger\hat{a}$ is the number operator, since it can be shown that

$$\hat{a}|n\rangle = n^{1/2}|n-1\rangle \quad \hat{a}^\dagger|n\rangle = (n+1)^{1/2}|n+1\rangle \quad (1.34)$$

Based on the quantum theory of the simple harmonic oscillator, the quantum theory of wave fields can be applied to field of lattice waves with annihilation and creation operators for waves with individual polarization and wave vector. Thus

$$\hat{H} = \sum_k \hbar\omega\left(\hat{a}^\dagger\hat{a} + \frac{1}{2}\right) \quad (1.35)$$

A lattice wave of the form $\hat{u} = \hat{A}e^{i(\mathbf{k}\cdot\mathbf{r}-\omega t)}$ has the classical energy in the cavity that contains the field:

$$m\omega^2 \int \hat{u}^* \cdot \hat{u} \, d\mathbf{r}/V = m\omega^2 \hat{A}^* \cdot \hat{A} = m\omega^2 \hat{a}^* \cdot \mathbf{a} \quad (1.36)$$

Thus

$$\hat{A} = \left(\frac{\hbar}{2m\omega^2}\right)^{1/2} (\mathbf{e_k}\hat{a}_\mathbf{k}e^{i\mathbf{k}\cdot\mathbf{r}} + \mathbf{e_k^*}\hat{a}_\mathbf{-k}^\dagger e^{-\mathbf{k}\cdot\mathbf{r}})e^{-i\omega t} \qquad (1.37)$$

This procedure for identifying the amplitude operator with annihilation and creation operators is often known as energy normalization. The example here is trivial; it really comes into its own for hybrids in extended cavities, as we will see in later chapters.

2
Optical Modes

2.1 Introduction

The description of acoustic vibrations in homogenous solids is well established in both continuum and atomic models. Continuum theory, dating back to Cauchy's work (1822), is firmly rooted in classical elasticity theory, which contains the mathematical definitions of stress and strain and their mutual relation in terms of elastic constants (see Chapter 1). However, solids consist of atoms and it is not evident that a continuum theory, which explicitly ignores the discrete nature of atoms, can describe lattice vibrations properly, especially in nanostructures which by their very nature introduce inhomogeneity in the form of interfaces between disparate materials. Microscopic lattice dynamics is, in principle, exact, but extremely computer-intensive and, moreover, produces results that lack the intuitive clarity that allows useful generalizations to be made. On the other hand, there exist transparent atomic models that are persuasive and have the useful property of being analytic. Such a model is Keating's valence-bond model (1966), which we will have cause to describe in some detail later in this chapter. Prima facie it would seem that continuum models would be redundant, but this is by no means the case. In fact, it can be demonstrated that a formalism in terms of envelope functions exists that exactly and uniquely describes atomic motion (Krumhansl 1965; Kunin 1982); on condition that the envelope functions are restricted to wave vectors within the first Brillouin zone. Thus, continuum theory (strictly quasi-continuum theory) is, in principle, as accurate as microscopic theory. A brief account can be found in Ridley (2009).

While acoustic waves are well described by either continuum or atomic theory, the same is not the case for optical waves. The classic theory of optical modes by Born and Huang (1954) is restricted to long wavelengths and does not describe lattice dispersion (the variation of frequency with wave vector), unlike the case for acoustic modes. In describing waves in nanostructures it is essential to identify appropriate boundary conditions at an interface. The long-wavelength model of optical modes in polar material can provide electromagnetic boundary conditions, but there are also mechanical boundary conditions. These involve the continuity of amplitude, force, and stress, and stress is related to dispersion.

It was clear that the case of optical phonons in nanostructures presented a problem. The displacement in an optical vibration involving two ions of different masses is different from that of an acoustic vibration. In the latter case, the energy of vibration is related to the displacement of the centre of gravity (**U**), whereas in the optical case it is related to the difference of displacement of the two ions (**u**). An optical strain can be defined by $S^{op}_{\lambda\mu} = \partial u_\lambda/\partial x_\mu$, but the question remained concerning the relation to the mechanical stress $T^{op}_{\alpha\beta}$. An acoustic-like Hooke's law relation could be assumed, that is $T^{op}_{\alpha\beta} = \sum_{\lambda\mu} c^{op}_{\alpha\beta\lambda\mu} S^{op}_{\lambda\mu}$, but what were the elastic constants? Could they be the same as the acoustic ones? After all, the bonds between the atoms were the same.

Focus on the atomic bonds provided the answer, either via the intense numerical methods of lattice dynamics involving sophisticated models of atomic bonds or via analytic approaches using simple models such as the valence-bond model of Keating (1966) that was applied to acoustic modes. Keating's model suggested that an application to optical modes could provide valuable insights into optical mode elasticity, and this was taken up by Foreman and Ridley (1999). A full account of this is overdue and is presented in what follows. It turns out that the optical elastic constants are not equal to the acoustic ones, nor do they have the same symmetry, and, moreover, they depend on ionic mass, making the mechanical boundary conditions significantly more complex than the acoustic ones.

We limit attention in this chapter to the non-polar diamond lattice in order to establish the mechanical boundary conditions; polar elements can be added later (Chapter 3).

2.2 Microscopic Theory of the Diamond Lattice

The basic translational unit of the diamond lattice is a cube of side equal to the lattice constant. It contains four unit cells, each of side a, and each containing two atoms. Each atom is bound to its four nearest neighbours by valence bonds of length $\sqrt{3}a$, directed to the corners of a tetrahedron (Fig. 2.1). When the two atoms vibrate together in the same direction, we have an acoustic wave. When the two atoms vibrate against one another, we have an optical wave.

We will assume that the valence forces that we need to consider are short range to the extent that atom A experiences forces associated with its four nearest neighbours (one of which includes atom B) plus forces associated with the 12 second-nearest neighbours. We thus extract a particular pattern of forces that is repeatable throughout the lattice. The four nearest neighbours of A reside on the B sublattice and will be denoted by a bar over its number, viz: $\bar{1}, \bar{2}, \bar{3}, \bar{4}$ with $\bar{1}$ denoting atom B. A number without an overbar denotes an atom on the A sublattice. Thus, the vector $X^{\bar{1}1}$ denotes the bond between $\bar{1}$ and 1 in the undisturbed lattice.

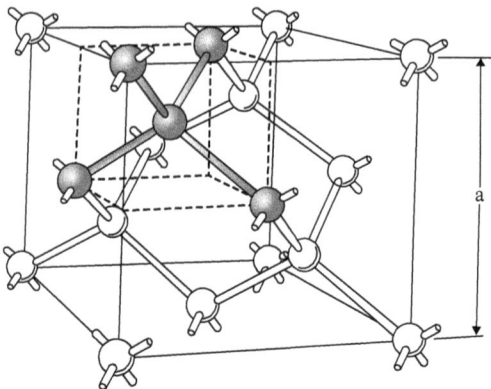

Figure 2.1 *The diamond lattice (from W. Shockley, Electrons and Holes in Semiconductors. Wiley/Van Nostrand Reinhold, New Jersey, 1950).*

Table 2.1 *Components of the bond vectors of nearest neighbours in units of a.*

	X	Y	Z
$X^{\bar{1}1}$	−1	−1	−1
$X^{\bar{2}1}$	−1	1	1
$X^{\bar{3}1}$	1	1	−1
$X^{\bar{4}1}$	1	−1	1

Taking atom A as the origin, we see that the components of $X^{\bar{1}1}$ are $a(-1,-1,-1)$ where a is the length of the side of the unit cell. The components of the vectors depicting the other neighbours of A can also be found. Table 2.1 summarizes the results.

We adopt the valence-bond model of Keating (1966). This model assumes that the potential energy of the deformed lattice is adequately described in terms of two force constants and quadratic terms involving the displacement of the atoms:

$$\Phi = \frac{1}{8a^2}\left\{\alpha\sum_{i=1}^{4}\Delta(\mathbf{x}^{\bar{i}0}.\mathbf{x}^{\bar{i}0})^2 + \beta^A\sum_{i,j>i}\Delta(\mathbf{x}^{\bar{i}0}.\mathbf{x}^{0\bar{j}})^2 + \beta^B\sum_{i,k\neq i}\Delta(\mathbf{x}^{0\bar{i}}.\mathbf{x}^{\bar{i}k})^2\right\} \quad (2.1)$$

The first term involves the central force constant α and the displacement of A relative to its nearest neighbours. The second term is over pairs of nearest neighbours and β^A is the relevant bond-bending force constant. The third term is over nearest neighbours (i) and the second-nearest neighbours (k) with β^B the relevant force constant. We follow Martin (1970) in assuming that β^A and β^B have, in general, different magnitudes. This allows us to apply our analysis to the case of the zinc blende lattice (without its polar property). We note that $\mathbf{x}^{\bar{i}0} = \mathbf{x}(\bar{i})-\mathbf{x}^A(0)$

and $\mathbf{x} = \mathbf{u} + \mathbf{X}$, where $\mathbf{u}$ is a small displacement from the atom's equilibrium position so that to lowest order:

$$\Delta(\mathbf{x}^p \cdot \mathbf{x}^q) = \mathbf{u}^p \cdot \mathbf{X}^q + \mathbf{u}^q \cdot \mathbf{X}^p$$

The equation of motion for atom A is then obtained from

$$\omega^2 m^A u_\alpha^A = \partial \Phi / \partial u_\alpha^A = \frac{1}{4a^2} \left\{ 4\alpha \sum_{i=1}^{4} u_\lambda^{\bar{i}0} X_\lambda^{\bar{i}0} X_\alpha^{\bar{i}0} + \beta^A \sum_{i,j \neq i} (u_\lambda^{\bar{i}0} X_\lambda^{0j} \right.$$

$$\left. + u_\lambda^{0j} X_\lambda^{\bar{i}0})(-X_\alpha^{0j} + X_\alpha^{\bar{i}0}) + \beta^B \sum_{i,k \neq i} (u_\lambda^{0i} X_\lambda^{k\bar{i}} - u_\lambda^{k\bar{i}} X_\lambda^{\bar{i}0}) X_\alpha^{k\bar{i}} \right\} \quad (2.2)$$

Note that in this notation repeated subscripts imply sums. Thus, for example, $u_\lambda X_\lambda = u_x X_x + u_y X_y + u_z X_z$. We are faced with summations over the bond vectors. As a first step it is useful to remove the restrictions on j and k. This produces extra terms that can be incorporated into the sum associated with the central force. We obtain

$$\omega^2 m^A u_\alpha^A = \frac{1}{4a^2} \left\{ 4(\alpha - \beta) \sum_{i=1}^{4} u_\lambda^{\bar{i}0} X_\lambda^{\bar{i}0} X_\alpha^{\bar{i}0} + \beta^A \sum_{i,j} (u_\lambda^{0j} X_\lambda^{\bar{i}0})(X_\alpha^{\bar{j}0} + X_\alpha^{\bar{i}0}) \right.$$

$$\left. + \beta^B \sum_{i,k} (u_\lambda^{0i} X_\lambda^{k\bar{i}} - u_\lambda^{k\bar{i}} X_\lambda^{\bar{i}0}) X_\alpha^{k\bar{i}} \right\} \quad (2.3)$$

where $\beta = \frac{1}{2}(\beta^A + \beta^B)$. We can now use the identities derived from Table 2.1 and similar relations for the second-nearest neighbours, namely:

$$\sum_i X_\alpha^{\bar{i}0} = 0 \quad (2.4)$$

$$\sum_i X_\alpha^{\bar{i}0} X_\beta^{\bar{i}0} = 4a^2 \delta_{\alpha\beta} \quad (2.5)$$

$$\sum_i X_\alpha^{\bar{i}0} X_\beta^{\bar{i}0} X_\gamma^{\bar{i}0} = 4a^3 |\varepsilon_{\alpha\beta\gamma}| \quad (2.6)$$

$$\sum_i X_\alpha^{\bar{i}0} X_\beta^{\bar{i}0} X_\sigma^{\bar{i}0} X_\tau^{\bar{i}0} = 4a^4 \Delta_{\alpha\beta\sigma\tau} \quad (2.7)$$

Optical Modes

The symbols ε, Δ are defined below. Applying equation (2.5) leads to

$$\omega^2 m^A u^A = \frac{1}{4a^2}\left\{4(\alpha-\beta)\sum_{i=1}^{4} u_\lambda^{\bar{i}0} X_\lambda^{\bar{i}0} X_\alpha^{\bar{i}0} + 8a^2\beta\sum_{i} u_\lambda^{0\bar{i}} - \beta^B \sum_{i,k} u_\lambda^{ki} X_\lambda^{\bar{i}0} X_\alpha^{ki}\right\} \quad (2.8)$$

We now make a bridge to the continuum by performing a second-order Taylor-series expansion about A:

$$u_\lambda^{\bar{i}0} = u_\lambda^B(\bar{i}) - u_\lambda^A(0) = u_\lambda^B + X_\sigma^{\bar{i}0} u_{\lambda,\sigma}^B + \frac{1}{2} X_\sigma^{\bar{i}0} X_\tau^{\bar{i}0} u_{\lambda,\sigma\tau}^B - u_\lambda^A \quad (2.9)$$

where we have used the notation $u_{\lambda,\sigma} = \partial u_\lambda/\partial x_\sigma$. As regards a similar expansion for u_λ^{ki}, we note to zero order $u_\lambda^{ki} = u_\lambda^A - u_\lambda^B = u_\lambda$, which is the optical displacement. Then

$$u_\lambda^{ki} = u_\lambda + X_\sigma^{\bar{i}0} u_{\lambda,\sigma} + X_\sigma^{ki} u_{\lambda,\sigma}^A + \frac{1}{2}(X_\sigma^{ki} X_\tau^{ki} + X_\sigma^{ki} X_\tau^{\bar{i}0} + X_\sigma^{\bar{i}0} X_\tau^{ki}) u_{\lambda,\sigma\tau}^A \quad (2.10)$$

Substituting into equation (2.8) we can evaluate the sums explicitly by using equations (2.6) and (2.7). In equation (2.6) $\varepsilon_{\alpha\beta\gamma}$ is the permutation operator which is $+1$ if all the subscripts are different and in cyclic order (e.g. xyz, yzx, zxy), and -1 if all the subscripts are different but are not in cyclic order, and zero otherwise. In our case the order does not matter, so we use the modulus. In equation (2.7) the operator $\Delta_{\alpha\beta\sigma\tau}$ is zero unless the subscripts come in matched pairs, in which case it is unity. It may be expressed as follows:

$$\Delta_{\alpha\beta\sigma\tau} = \delta_{\alpha\beta}\delta_{\sigma\tau} + (1-\delta_{\alpha\beta})(\delta_{\alpha\sigma}\delta_{\beta\tau} + \delta_{\alpha\tau}\delta_{\beta\sigma}) \quad (2.11)$$

If we replace $X_\beta^{\bar{i}0}$ with X_β^{ki} in equations (2.4–2.7), everything remains the same except that equation (2.7) changes sign.

We are now able to write down the explicit equation of motion for atom A:

$$\omega^2 m^A u_\alpha^A = 4(\alpha+\beta)u_\alpha - 4a(\alpha-\beta)|\varepsilon_{\alpha\beta\gamma}|u_{\lambda,\sigma}^B - 2a^2[(\alpha-\beta)\Delta_{\alpha\beta\sigma\tau}+2\beta\delta_{\alpha\beta}\delta_{\sigma\tau}]u_{\lambda,\sigma\tau}^B$$
$$- 2a^2\beta^B(\delta_{\alpha\sigma}\delta_{\lambda\tau} + \delta_{\alpha\tau}\delta_{\lambda\sigma})u_{\lambda,\sigma\tau}^A \quad (2.12)$$

The equation of motion for B is obtained simply by interchanging labels A and B and changing the sign of the linear terms:

$$\omega^2 m^B u_\alpha^B = -4(\alpha+\beta)u_\alpha + 4a(\alpha-\beta)|\varepsilon_{\alpha\beta\gamma}|u_{\lambda,\sigma}^A - 2a^2[(\alpha-\beta)\Delta_{\alpha\beta\sigma\tau}+2\beta\delta_{\alpha\beta}\delta_{\sigma\tau}]u_{\lambda,\sigma\tau}^A$$
$$- 2a^2\beta^A(\delta_{\alpha\sigma}\delta_{\lambda\tau} + \delta_{\alpha\tau}\delta_{\lambda\sigma})u_{\lambda,\sigma\tau}^B \quad (2.13)$$

Notice that we have assumed that atoms A and B have, in general, different masses.

2.3 Decoupled Acoustic and Optical Equations

The next step is to convert these equations into equations that describe acoustic and optical waves. The acoustic and optical displacements are associated with the displacement of the centre of gravity in the acoustic case, and with the relative displacement of the two sublattices in the optical case:

$$MU = m^A u^A + m^B u^B$$
$$M = m^A + m^B \tag{2.14}$$
$$u = u^A - u^B$$

Adding the equations for the individual atoms gives the acoustic equation:

$$\omega^2 MU_\alpha = 4a(\alpha - \beta)|\varepsilon_{\alpha\beta\gamma}|u_{\lambda,\sigma} - 2a^2[(\alpha - \beta)\Delta_{\alpha\beta\sigma\tau} + 2\beta\delta_{\alpha\beta}\delta_{\sigma\tau}](u^A_{\lambda,\sigma\tau} + u^B_{\lambda,\sigma\tau})$$
$$- 2a^2(\delta_{\alpha\sigma}\delta_{\lambda\tau} + \delta_{\alpha\tau}\delta_{\lambda\sigma})(\beta^B u^A_{\lambda,\sigma\tau} + \beta^A u^B_{\lambda,\sigma\tau}) \tag{2.15}$$

Multiplying equation (2.12) by m^B/M and equation (2.13) by m^A/M and subtracting gives the optical equation:

$$\omega^2 \mu u_\alpha = 4(\alpha + \beta)u_\alpha - 4a(\alpha - \beta)|\varepsilon_{\alpha\beta\gamma}|(r^A u^A_{\lambda,\sigma} + r^B u^B_{\lambda,\sigma})$$
$$+ 2a^2[(\alpha - \beta)\Delta_{\alpha\beta\sigma\tau} + 2\beta\delta_{\alpha\beta}\delta_{\sigma\tau}](r^A u^A_{\lambda,\sigma\tau} - r^B u^B_{\lambda,\sigma\tau}) \tag{2.16}$$
$$- 2a^2(\delta_{\alpha\sigma}\delta_{\lambda\tau} + \delta_{\alpha\tau}\delta_{\lambda\sigma})(\beta^B r^B u^A_{\lambda,\sigma\tau} + \beta^A r^A u^B_{\lambda,\sigma\tau})$$

where $\mu = Mr^A r^B$ is the reduced mass, and $r^A = m^A/M, r^B = m^B/M$.

In order to replace the displacements of the individual atoms in terms of acoustic and optical displacements it is useful to be aware of the relationships collected together in Table 2.2.

Table 2.2 *Useful relations.*

$$u^A_\lambda = U_\lambda + r^B u_\lambda$$
$$u^B_\lambda = U_\lambda - r^A u_\lambda$$
$$\beta = \frac{1}{2}(\beta^A + \beta^B)$$
$$\Delta\beta = \frac{1}{2}(\beta^A - \beta^B)$$
$$r^A + r^B = 1$$
$$r = \frac{1}{2}(r^A - r^B)$$

30 Optical Modes

The terms involving u^A and u^B can be expressed in terms of acoustic and optical displacements:

$$u^A_{\lambda,\sigma\tau} + u^B_{\lambda,\sigma\tau} = 2U_{\lambda,\sigma\tau} - 2(ru_\lambda)_{,\sigma\tau}$$

$$\beta^B u^A_{\lambda,\sigma\tau} + \beta^A u^B_{\lambda,\sigma\tau} = 2\beta U_{\lambda,\sigma\tau} - 2\beta(ru_\lambda)_{,\sigma\tau} - \Delta\beta u_{\lambda,\sigma\tau}$$

$$r^A u^A_{\lambda,\sigma} + r^B u^B_{\lambda,\sigma} = U_{\lambda,\sigma} - r_{,\sigma} u_\lambda \qquad (2.17)$$

$$r^A u^A_{\lambda,\sigma\tau} + r^B u^B_{\lambda,\sigma\tau} = 2rU_{\lambda,\sigma\tau} - 2rr_{,\sigma\tau}u_\lambda + [(\mu/M)u_{\lambda,\sigma}]_{,\tau} + [(\mu/M)u_{\lambda,\tau}]_{,\sigma}$$

$$\beta^B r^B u^A_{\lambda,\sigma\tau} + \beta^A r^A u^B_{\lambda,\sigma\tau} = \beta[u_{\lambda,\sigma\tau} - (r^A u^A_{\lambda,\sigma\tau} + r^B u^B_{\lambda,\sigma\tau})] - \Delta\beta[U_{\lambda,\sigma\tau}$$
$$- r_{,\sigma\tau}u_\lambda - (ru_{\lambda,\sigma})_{,\tau} - (ru_{\lambda,\tau})_{,\sigma}]$$

Note that these expressions include the possibility of a spatial variation of the mass factors. In homogenous material such spatial variations vanish, but in nanostructures they have to be retained as they will contribute to the boundary conditions at an interface. Substitution into equations (2.15) and (2.16) yields equations of motion purely in terms of acoustic and optical displacements:

$$\omega^2 M U_\alpha = 4a(\alpha - \beta)|\varepsilon_{\alpha\beta\gamma}|u_{\lambda,\sigma} - 4a^2[(\alpha - \beta)\Delta_{\alpha\beta\sigma\tau} + 2\beta\delta_{\alpha\beta}\delta_{\sigma\tau}](U_{\lambda,\sigma\tau} - (ru_\lambda)_{,\sigma\tau})$$
$$- 4a^2(\delta_{\alpha\sigma}\delta_{\lambda\tau} + \delta_{\alpha\tau}\delta_{\lambda\sigma})(\beta U_{\lambda,\sigma\tau} - \beta(ru_\lambda)_{,\sigma\tau} - (\Delta\beta/2)u_{\lambda,\sigma\tau}) \qquad (2.18)$$

$$\omega^2 \mu u_\alpha = 4(\alpha + \beta)u_\alpha - 4a(\alpha - \beta)|\varepsilon_{\alpha\beta\gamma}|(U_{\lambda,\sigma} - r_{,\sigma}u_\lambda)$$
$$+ 2a^2[(\alpha - \beta)\Delta_{\alpha\beta\sigma\tau} + 2\beta\delta_{\alpha\beta}\delta_{\sigma\tau}](2rU_{\lambda,\sigma\tau} - 2rr_{,\sigma\tau}u_\lambda + [(\mu/M)u_{\lambda,\sigma}]_{,\tau}$$
$$+ [(\mu/M)u_{\lambda,\tau}]_{,\sigma}) - 2a^2(\delta_{\alpha\sigma}\delta_{\lambda\tau} + \delta_{\alpha\tau}\delta_{\lambda\sigma})$$
$$\times \{\beta[u_{\lambda,\sigma\tau} - (r^A u^A_{\lambda,\sigma\tau} + r^B u^B_{\lambda,\sigma\tau})] - \Delta\beta[U_{\lambda,\sigma\tau} - r_{,\sigma\tau}u_\lambda - (ru_{\lambda,\sigma})_{,\tau} - (ru_{\lambda,\tau})_{,\sigma}]\}$$
$$(2.19)$$

These results show that acoustic and optical waves are interdependent in general. They can be decoupled, however, in a long-wavelength approximation in which the optical frequency is taken to be very much larger than the acoustic frequency at sufficiently long wavelengths. Thus, in the optical equation (2.19), the frequency can be taken to be vanishingly small in the acoustic case. Equation (2.19) can be satisfied if the leading terms are related according to

$$u_\alpha = a\frac{(\alpha - \beta)}{(\alpha + \beta)}|\varepsilon_{\alpha\beta\gamma}|U_{\lambda,\sigma} \qquad (2.20)$$

Substitution into equation (2.18) and dispensing with terms higher than second order yields the uncoupled acoustic equation. A similar argument leads to the uncoupled optical equation. Thus, taking the frequency to be very large in equation (2.18) the acoustic equation can be satisfied by the leading terms being related according to

$$U_\alpha = 4a \frac{(\alpha - \beta)}{\omega^2 M} |\varepsilon_{\alpha\beta\gamma}| u_{\lambda,\sigma} \qquad (2.21)$$

Substitution into equation (2.19) and, once again, dispensing with terms of higher order than second produces the uncoupled optical equation. The uncoupled equations are

$$\omega^2 M U_\alpha = 4a^2 \left[\frac{(\alpha - \beta)^2}{\alpha + \beta} |\varepsilon_{\alpha\beta\gamma}||\varepsilon_{\sigma\tau\mu}| - [(\alpha - \beta)\Delta_{\alpha\beta\sigma\tau} + \beta(2\delta_{\alpha\beta}\delta_{\sigma\tau} \right.$$

$$\left. + \delta_{\alpha\sigma}\delta_{\beta\tau} + \delta_{\alpha\tau}\delta_{\beta\sigma})] \right] U_{\beta,\sigma\tau} \qquad (2.22)$$

$$\omega^2 \mu u_\alpha = 4(\alpha + \beta) u_\lambda - 16a \frac{(\alpha - \beta)^2}{\omega^2 M} |\varepsilon_{\alpha\beta\gamma}||\varepsilon_{\sigma\tau\mu}| u_{\lambda,\sigma\tau} + 4a(\alpha - \beta)|\varepsilon_{\alpha\beta\gamma}| r_{,\sigma} u_\lambda$$

$$+ 2a^2 ([(\alpha - \beta)\Delta_{\alpha\beta\sigma\tau} + \beta(2\delta_{\alpha\beta}\delta_{\sigma\tau} + \delta_{\alpha\sigma}\delta_{\beta\tau} + \delta_{\alpha\tau}\delta_{\beta\sigma})](-2rr_{,\sigma\tau}$$

$$+ [(\mu/M) u_{\lambda,\sigma}]_{,\tau} + [(\mu/M) u_{\lambda,\tau}]_{,\sigma}) - 2a^2 (\delta_{\alpha\sigma}\delta_{\beta\tau} + \delta_{\alpha\tau}\delta_{\beta\sigma})\{\beta u_{\lambda,\sigma\tau}$$

$$+ \Delta\beta[r_{,\sigma\tau} u_\lambda + (r u_{\lambda,\sigma})_{,\tau} + (r u_{\lambda,\tau})_{,\sigma}]\} \qquad (2.23)$$

It is worth noticing that any inhomogeneity causing the mass terms to vary in space does not affect the acoustic equation, but only the optical equation.

These equations are splendid in their generality, but precisely that property makes them somewhat indigestible. The connection with the classical elastic constants may become more transparent by focusing on a particular component of the displacement vector in each case. It will also be necessary to divide throughout by the volume of the unit cell in order to convert to density. The lattice cube has a side equal to 4a and, consequently, a volume of $64a^3$. There being four unit cells in the unit cube, the volume of the unit cell is $16a^3$.

As a preliminary, a clarification of the meaning of the product of the permutation operators would be useful. The components of equation (2.23) are

$$u_\lambda = C |\varepsilon_{\alpha\beta\gamma}| U_{\lambda,\sigma} \qquad (2.24)$$

$$u_x = C(U_{y,z} + U_{z,y}) \quad u_y = C(U_{z,x} + U_{x,z}) \quad u_z = C(U_{x,y} + U_{y,x})$$

In equation (2.22), therefore, for the x component:

$$|\varepsilon_{\alpha\lambda\gamma}||\varepsilon_{\sigma\beta\mu}| U_{\beta,\sigma\tau} = (1 - \delta_{\alpha\lambda})(\delta_{\alpha\beta}\delta_{\lambda\mu} + \delta_{\alpha\mu}\delta_{\beta\lambda}) U_{\beta,\sigma\tau} \qquad (2.25)$$

The acoustic equation can be written as

$$\omega^2 \rho U_\alpha = -(c_{\alpha\lambda\beta\mu} U_{\beta,\mu})_{,\lambda} = -c_{\alpha\lambda\beta\mu} U_{\beta,\mu\lambda} \tag{2.26}$$

As discussed in Section 1.5, Chapter 1, we assume that the elastic constants in the adjacent materials are the same. The explicit expression for the x component is

$$\omega^2 \rho U_x = -\left(c_{11} U_{x,xx} + c_{12}(U_{y,y} + U_{z,z})_{,x} + c_{44}[(U_{x,y} + U_{y,x})_{,y} + (U_{x,z} + U_{z,x})_{,z}]\right) \tag{2.27}$$

where

$$c_{11} = \frac{(\alpha + 3\beta)}{4a}, \quad c_{12} = \frac{(\alpha - \beta)}{4a}, \quad c_{44} = \frac{\alpha\beta}{a(\alpha + \beta)} \tag{2.28}$$

These are the relationships originally obtained by Keating. Here they have been shown to hold good for inhomogeneous material. Also note that the classic condition for the material to be elastically isotropic, $c_{11} - c_{12} - 2c_{44} = 0$, requires the extreme condition $\beta = 0$, in which case $c_{44} = 0$. Real crystals are always anisotropic. An assumption that a crystal is isotropic leads to very useful simplifying features, and it will be used frequently in forthcoming chapters, though with the condition that c_{44} is not zero.

The corresponding relations for the optical case are not obtained so straightforwardly. We first limit our attention to the case of a homogeneous crystal. This gets rid of all the mass dependences on space. The solution embodied in equation (2.25) is directly applicable to the optical case, with u replacing U. We can write the optical equation as follows:

$$\omega^2 \bar{\rho} u_x = c_{xx} u_x + c_{11}^{op} u_{x,xx} + c_{12}^{op}(u_{y,y} + u_{z,z})_{,x} + [c_{a44}^{op} u_{x,y} + c_{b44}^{op} u_{y,x}]_{,y} + [c_{a44}^{op} u_{x,z} + c_{b44}^{op} u_{z,x}]_{,z} \tag{2.29}$$

where

$$c_{xx} = \frac{\alpha + \beta}{4a^3}$$

$$c_{11}^{op} = \frac{\mu}{M}\frac{\alpha + 3\beta}{4a} - \frac{\beta'}{4a}$$

$$c_{12}^{op} = \frac{\mu}{M}\frac{\alpha}{4a} - \frac{\beta'}{8a} \tag{2.30}$$

$$c_{a44}^{op} = \frac{\mu}{M}\frac{\alpha + \beta}{4a} - \frac{(\alpha - \beta)^2}{a\omega^2 M}$$

$$c_{b44}^{op} = \frac{\mu}{M}\frac{\alpha}{4a} - \frac{(\alpha - \beta)^2}{a\omega^2 M} - \frac{\beta'}{8a}$$

where $\beta' = \beta + 2r\Delta\beta$. There are now two shear moduli instead of one in the acoustic case, indicating the lack of rotational invariance in the elasticity of the optical modes. Moreover, the shear moduli are frequency dependent, but a dependency that is only as strong as the bulk dispersion, and will be small at the longest wavelengths. It may be noted that the frequency-dependent factor can be written in terms of the acoustic elastic constant c_{12}:

$$\frac{(\alpha-\beta)^2}{a\omega^2 M} = \frac{4(\alpha-\beta)}{\omega^2 M} c_{12} = \frac{\mu\alpha}{M\,4a} \times \frac{16a(\alpha-\beta)}{a\omega^2\mu} c_{12} < \frac{\mu\alpha}{M\,4a} \times \frac{16a}{\omega^2\mu} \quad (2.31)$$

In the case of GaAs $16a/\omega^2\mu \approx 0.07$ which is, indeed, small.

2.4 Velocities

A comment on the meaning of the term dispersion is in order. In classical physics dispersion refers to the variation of velocity with wavelength and is applicable to acoustic waves. Dispersion, used in the context of optical modes, refers to the variation of frequency with wavelength, and to this property we now turn.

Equation 2.29 can be expressed as follows:

$$\omega^2 \bar{\rho} u_x = c_{xx} u_x + c^{op}_{a44}(u_{x,xx} + u_{y,yy} + u_{z,zz}) + (c^{op}_{12} + c^{op}_{b44})(u_{x,x} + u_{y,y} + y_{z,z})_{,x} + c^{op}_0 u_{x,xx}$$

$$c^{op}_0 = c^{op}_{11} - c^{op}_{12} - c^{op}_{a44} - c^{op}_{b44} \quad (2.32)$$

If we introduce the frequency ω_0, such that $\omega_0^2 \bar{\rho} = c_{xx}$, and generalize the equation, we get

$$(\omega^2 - \omega_0^2)\bar{\rho}\mathbf{u} = c^{op}_{a44}\nabla^2 \mathbf{u} + (c^{op}_{12} + c^{op}_{b44})\nabla\nabla.\mathbf{u} + c^{op}_0(\hat{\mathbf{x}} u_{x,xx} + \hat{\mathbf{y}} u_{y,yy} + \hat{\mathbf{z}} u_{z,zz}) \quad (2.33)$$

where $\hat{\mathbf{x}}$, etc. are unit vectors. For a travelling wave: $\mathbf{u} = \mathbf{a} e^{i(\mathbf{k}.\mathbf{r}-\omega t)}$, having a frequency obeying

$$\omega^2 = \omega_0^2 - v^2 k^2 \quad (2.34)$$

the equation becomes

$$v^2 k^2 \bar{\rho}\mathbf{a} = c^{op}_{a44} k^2 \mathbf{a} + (c^{op}_{12} + c^{op}_{b44})\mathbf{k}(\mathbf{k}.\mathbf{a}) + c^{op}_0(\hat{\mathbf{x}} k_x^2 a_x + \hat{\mathbf{y}} k_y^2 a_y + \hat{\mathbf{z}} k_z^2 a_z) \quad (2.35)$$

If k is expressed in terms of its direction cosines with respect to the principal axes of the crystal, the equation can be written as

$$v^2 \bar{\rho} a_x = [c^{op}_{a44} + \alpha^2(c^{op}_{12} + c^{op}_{b44} + c^{op}_0)]a_x + (c^{op}_{12} + c^{op}_{b44})\alpha\beta a_y + (c^{op}_{12} + c^{op}_{b44})\alpha\gamma a_z \quad (2.36)$$

Table 2.3 $\bar{\rho}v^2$.

	[100]	[110]	[111]
LO	$c_{12}^{op} + c_{a44}^{op} + c_{b44}^{op} + c_o^{op}$	$c_{12}^{op} + c_{a44}^{op} + c_{b44}^{op} + c_o^{op}/2$	$c_{12}^{op} + c_{a44}^{op} + c_{b44}^{op} + c_o^{op}/3$
TO$_1$	c_{a44}^{op}	$c_{a44}^{op} + c_o^{op}/2$	$c_{a44}^{op} + c_o^{op}/3$
TO$_2$	c_{a44}^{op}	c_{a44}^{op}	$c_{a44}^{op} + c_o^{op}/3$

with corresponding expressions for the y and z components. Solutions for the LO and TO velocities are shown in Table 2.3.

2.5 Isotropy

The condition for isotropy becomes $c_{11}^{op} - c_{12}^{op} - c_{a44}^{op} - c_{b44}^{op} = 0$. Noting that in the long-wavelength limit $\omega^2 = 4(\alpha + \beta)/\mu$, we see that the condition is once more satisfied by $\beta = 0$. If this were true, then $c_{11}^{op} = c_{12}^{op}, c_{a44}^{op} = c_{b44}^{op} = 0$. No material involved in nanostructures has this property but such is the simplification to be gained by assuming the crystal is isotropic, we need to examine what is the magnitude of error involved. Taking GaAs as the paradigmatic nanostructure crystal, we can compute the ratio β/α using the magnitudes of Table 1.2. The ratio c_{12}/c_{11} gives $\beta/\alpha = 0.16$ and c_{44}/c_{11} gives 0.17. Similar results are obtained for AlAs. Thus, the error involved assuming isotropy in these cases is no worse than 17%.

If the assumption of homogeneous, isotropic material is made, the acoustic and optical equations of motion can be written as

$$\omega^2 \rho \mathbf{U} = -c_{11} \nabla \nabla . \mathbf{U} + c_{44} \nabla \times \nabla \times \mathbf{U}$$
$$(\omega^2 - \omega_0^2) \bar{\rho} \mathbf{u} = c_{11}^{op} \nabla \nabla . \mathbf{u} - c_{a44}^{op} \nabla \times \nabla \times \mathbf{u}$$
(2.37)

In the derivation of these equations we have used the identity $\nabla^2 \mathbf{U} = \nabla \times \nabla \times \mathbf{U} - \nabla \nabla . \mathbf{U}$. Note that only one of the optical shear coefficients is effective in isotropic material. Moreover, longitudinally polarized ($\nabla \times \mathbf{u} = 0$) and transversely polarized ($\nabla . \mathbf{u} = 0$) modes now become distinct, a blessing for which to be truly thankful. Another useful result is that $c_{11}^{op}/\bar{\rho} \approx c_{11}/\rho, c_{44}^{op}/\bar{\rho} = c_{44}/\rho$, which makes the LO and TO velocities approximately equal to the LA and TA velocities.

2.6 Inhomogeneous System

If we consider a lattice-matched structure such as GaAs/AlAs, in which the interface can be defined by the atom common to both semiconductors, namely arsenic, then it would seem that the acoustic boundary conditions should apply. The displacement **u** of the arsenic atom is what it is, independent of GaAs and AlAs. In this sense, the continuity of displacement is guaranteed. Equally, the atom experiences no net force that would force it out of position. Since a net force is associated with the variation of stress, it means that the stress must be continuous across the interface. The connection rules are therefore just those of the acoustic case.

What is wrong with this argument is the assumption that the interface can be defined by the position of a particular atom. In the present case, the arsenic atom has bonds with gallium on one side and with aluminium on the other. The physical nature of the interface is therefore more diffuse, involving the variation of bond strengths and mass.

Because of the variation of mass, the situation for optical modes in the vicinity of a boundary is much more complex than it is for acoustic modes, even when the force constants themselves are assumed not to change. The full equation of motion is

$$\omega^2 \bar{\rho} u_\alpha = c_{\alpha\beta} u_\beta + (c_{\alpha\lambda\beta\mu} u_{\beta,\mu})_{,\lambda} \tag{2.38}$$

with

$$c_{\alpha\beta} = \frac{\alpha+\beta}{4a^3} + \frac{\alpha-\beta}{4a^2} |\varepsilon_{\alpha\beta\mu}| r_{,\mu} + \frac{1}{8a}[(\alpha-\beta)\Delta_{\alpha\beta\sigma\tau} + \beta(2\delta_{\alpha\beta}\delta_{\sigma\tau}$$
$$+ \delta_{\alpha\sigma}\delta_{\beta\tau} + \delta_{\alpha\tau}\delta_{\beta\sigma})] \left[\left(\frac{\mu}{M}\right)_{,\lambda\mu} + 2r_{,\lambda} r_{,\mu}\right] - \frac{\Delta\beta}{16a}(\delta_{\alpha\beta}\delta_{\lambda\mu} + \delta_{\alpha\mu}\delta_{\beta\lambda})r_{,\lambda\mu}$$
$$\tag{2.39}$$

$$c_{\alpha\lambda\beta\mu} = \frac{1}{4a}\frac{\mu}{M}\left([(\alpha-\beta)\Delta_{\alpha\beta\sigma\tau} + \beta(2\delta_{\alpha\beta}\delta_{\sigma\tau} + \delta_{\alpha\sigma}\delta_{\beta\tau} + \delta_{\alpha\tau}\delta_{\beta\sigma})]\right)$$
$$- \frac{\beta'}{8a}(\delta_{\alpha\beta}\delta_{\lambda\mu} + \delta_{\alpha\mu}\delta_{\beta\lambda}) - \frac{(\alpha-\beta)^2}{a\omega^2 M}(1-\delta_{\alpha\lambda})(\delta_{\alpha\beta}\delta_{\lambda\mu} + \delta_{\alpha\mu}\delta_{\beta\lambda}) \tag{2.40}$$

For a crystal inhomogeneous in the z direction, the non-zero elements are

$$c_{zz} = \frac{\alpha+\beta}{4a_3} + \frac{\alpha+3\beta}{8a}\left[\left(\frac{\mu}{M}\right)_{,zz} + 2(r_{,z})^2\right]$$

$$c_{xx} = c_{yy} = \frac{\alpha+\beta}{4a_3} + \frac{\alpha+\beta}{8a}\left[\left(\frac{\mu}{M}\right)_{,zz} + 2(r_{,z})^2\right] \tag{2.41}$$

$$c_{xy} = c_{yx} = \frac{\alpha-\beta}{4a^2} r_{,z}$$

We note that the assumption of an abrupt interface will imply delta-function terms associated with the ratio of the reduced to the total mass and with the difference-in-mass factors. In establishing connection rules for acoustic modes only the amplitude and stress had to be continuous. In establishing connection rules for optical modes the variation of the force factors and the mass factors along with amplitude and stress have to be taken into account. As in the acoustic case, it is often assumed that the force constants do not differ appreciably across a lattice-matched interface, an approximation referred to as the mass approximation.

The boundary conditions will be considered in Chapter 4.

3
Polar Modes in Zinc Blende

3.1 Polar Elements

In Chapter 2, a continuum theory of long-wavelength non-polar acoustic and optical modes arising out of a microscopic model of the lattice dynamics was described. Here, we add the polar elements that occur in III-V compounds. The zinc blende lattice is cubic with tetrahedral bonding and two atoms per unit cell. The two atoms are different—for example, GaAs—and carry opposite electrical charges. As a consequence, acoustic strains (in the zinc blende case only shear strains) and optical displacements produce an electrical polarization of the lattice. In general, the electric displacement, $\mathbf{D}$, is determined by the electric field, E, plus the electric polarization due to an optical displacement, u, and due, via the piezoelectric effect, to a strain S_{ij}:

$$D_i = \sum_{ij} \varepsilon_{ij} E_j + \sum_j \frac{e^*_{ij}}{V_0} u_{ij} + \sum_{kl} e_{ikl} S_{kl} \tag{3.1}$$

where ε is the permittivity tensor, e^* is the ionic-charge tensor, V_0 is the volume of the unit cell, and e is the piezoelectric tensor (now no longer the permutation operator). We label strain by S. The restoring force, F, associated with the optical displacement now involves the electric field through the ionic charge:

$$F_i = \sum_j c^{op}_{ij} u_j - \sum_j e^*_{ij} E_j \tag{3.2}$$

Here, c^{op} is the restoring force tensor. The elastic stress tensor, $\mathbf{T}$, is related to strain and the electric field:

$$T_{ij} = \sum_{kl} c_{ijkl} S_{kl} - \sum_k e_{ijk} E_k \tag{3.3}$$

where c is the elastic constant tensor, different for acoustic and optical vibrations.

The symmetry of the cubic lattice allows some simplification. The permittivity is a scalar determined by the rapid response of the core electrons, so $\varepsilon_{ij} = \varepsilon_\infty \delta_{ij}$,

Hybrid Phonons in Nanostructures. First Edition. B.K. Ridley. © B.K. Ridley 2017.
Published in 2017 by Oxford University Press. DOI: 10.1093/acprof:oso/9780198788362.001.0001

also $e_{ij}^* = e^*\delta_{ij}$ and $c_{ij}^{op} = c^{op}\delta_{ij}$. There is only one non-zero piezoelectric coefficient, namely: $e_{123} = e_{213} = e_{312}$. In reduced notation: $11 = 1, 22 = 2, 33 = 3, 23 = 32 = 4, 13 = 31 = 5$, and $12 = 21 = 6$. Thus, the single piezoelectric coefficient is $e_{14} = e_{15} = e_{16}$ in the same way that the elastic constants are $c_{11} (= c_{22} = c_{33}), c_{12} (= c_{23} = c_{13})$, and c_{44} $(= c_{55} = c_{66})$. Incorporating these reductions into equations (3.1), (3.2), and (3.3), and writing the result out in full, we get

$$D_1 = \varepsilon_\infty E_1 + e^* u_1/V_0 + e_{14} S_4$$
$$D_2 = \varepsilon_\infty E_2 + e^* u_2/V_0 + e_{14} S_5$$
$$D_3 = \varepsilon_\infty E_3 + e^* u_3/V_0 + e_{14} S_6$$
$$F_j = c u_j^{opt} - e^* E_j \qquad (3.4)$$

Also, for acoustic modes (the optical mode stresses are treated in Chapter 4):

$$T_1 = c_{11} S_1 + c_{12}(S_2 + S_3)$$
$$T_2 = c_{11} S_2 + c_{12}(S_3 + S_1)$$
$$T_3 = c_{11} S_3 + c_{12}(S_1 + S_2)$$
$$T_4 = c_{44} S_4$$
$$T_5 = c_{44} S_5$$
$$T_6 = c_{44} S_6 \qquad (3.5a)$$

$$S_i = \frac{\partial U_i}{\partial x_i} \quad i = 1, 2, 3$$
$$S_4 = \frac{\partial U_y}{\partial z} + \frac{\partial U_z}{\partial y} \quad S_5 = \frac{\partial U_z}{\partial x} + \frac{\partial U_x}{\partial z} \quad S_6 = \frac{\partial U_x}{\partial y} + \frac{\partial U_y}{\partial x} \qquad (3.5b)$$

3.2 Polar Optical Modes

Optical displacements of atoms with opposite electric charges produce macroscopic electric fields associated with the polarization of the lattice. The relevant electric displacement, the sum of the electric polarization of the core electrons and the ionic polarization (dipole per unit volume), is given by

$$\mathbf{D} = \varepsilon(\omega)\mathbf{E} = \varepsilon_\infty \mathbf{E} + (e^*/V_0)\mathbf{u} \qquad (3.6)$$

where $\varepsilon(\omega), \varepsilon_\infty$ are the permittivities at the relevant frequency and at high frequencies respectively, e^* is the effective charge on the ion, and V_0 is the unit cell volume. The equation is coupled with Gauss's equation in the absence of free charge:

$$\nabla \cdot \mathbf{D} = 0 \tag{3.7}$$

From equation (3.6) we obtain the important relation between field and optical displacement:

$$\mathbf{E}(\omega) = \frac{e^*}{V_0[\varepsilon(\omega) - \varepsilon_\infty]} \mathbf{u}(\omega) \tag{3.8}$$

For a travelling wave, equation (3.7) becomes

$$\varepsilon(\omega)\mathbf{k} \cdot \mathbf{u} = 0 \tag{3.9}$$

This implies that for longitudinally polarized waves, where by definition $\mathbf{k} \cdot \mathbf{u} \neq 0$, the permittivity is zero. If ω_L denotes the frequency of a longitudinally polarized mode, then

$$\varepsilon(\omega_L) = 0 \tag{3.10}$$

On the other hand, $\mathbf{k} \cdot \mathbf{u} = 0$ for transversely polarized waves. This implies that the permittivity can be nonzero, but only for transverse waves.

We can find the magnitude of the ionic charge by considering the case of static polarization in the long-wavelength approximation, in which the equation of motion becomes

$$\omega^2 \bar{\rho} \mathbf{u} = c_{\alpha\alpha} \mathbf{u} - (e^*/V_0)\mathbf{E} \tag{3.11}$$

Here, $\bar{\rho} = \mu/V_0$ is the reduced-mass density. At zero frequency the permittivity is $\varepsilon(0) = \varepsilon_s$, the static permittivity, and the field is, from equation (3.8),

$$\mathbf{E}(0) = \frac{e^*}{V_0(\varepsilon_s - \varepsilon_\infty)} \mathbf{u}(0) \tag{3.12}$$

We may express the non-polar restoring force in terms of the equivalent non-polar frequency, thus

$$c_{\alpha\alpha} \mathbf{u} = \omega_T^2 \bar{\rho} \mathbf{u} \tag{3.13}$$

Substitution of equations (3.12) and (3.13) into equation (3.11) gives the ionic charge as

$$e^{*2} = \omega_T^2 \mu V_0 (\varepsilon_s - \varepsilon_\infty) \tag{3.14}$$

The equation of motion for an arbitrary frequency becomes

$$\omega^2 \bar{\rho}\mathbf{u} = \omega_T^2 \bar{\rho}\mathbf{u} - \frac{\omega_T^2 \mu(\varepsilon_s - \varepsilon_\infty)}{V_0[\varepsilon(\omega) - \varepsilon_\infty]}\mathbf{u} \qquad (3.15)$$

This provides us with the frequency dependence of the permittivity:

$$\varepsilon(\omega) = \frac{\omega^2 \varepsilon_\infty - \omega_T^2 \varepsilon_s}{\omega^2 - \omega_T} = \varepsilon_\infty \frac{\omega^2 - \omega_L^2}{\omega^2 - \omega_T^2} \qquad (3.16)$$

where we have introduced the frequency for the longitudinal modes:

$$\omega_L^2 = \omega_T^2 (\varepsilon_s/\varepsilon_\infty) \qquad (3.17)$$

This is known as the Lyddane–Sachs–Teller relation. The frequency dependence of the field from equation (3.8) becomes

$$\mathbf{E}(\omega) = -\frac{e^*}{V_0 \varepsilon_\infty} \frac{\omega^2 - \omega_T^2}{\omega_L^2 - \omega_T^2} \mathbf{u}(\omega) \qquad (3.18)$$

Gauss's equation becomes

$$\nabla \left(\frac{e^*}{V_0} \frac{\omega_L^2 - \omega^2}{\omega_L^2 - \omega_T^2} \mathbf{u}(\omega) \right) = 0 \qquad (3.19)$$

The attributions that have been given of the frequencies ω_L, ω_T can now be made clear. For a longitudinal wave we must have $\omega = \omega_L$; otherwise the wave is transverse. This includes the case of electromagnetic waves, which are all transverse.

3.2.1 Effects of Dispersion

Including dispersion (equation (2.29)), the equation of motion is

$$\omega^2 \bar{\rho} u_x = \omega_T^2 \bar{\rho} u_x - \frac{\omega_T^2 \bar{\rho}(\varepsilon_s - \varepsilon_\infty)}{[\varepsilon(\omega) - \varepsilon_\infty]} u_x$$
$$+ c_{11}^{op} u_{x,xx} + (c_{12}^{op} + c_{a44}^{op} + c_{b44}^{op})(u_{y,y} + u_{z,z})_{,x} - [\nabla \times c_{a44}^{op} \nabla \times \mathbf{u}]_{,x} \qquad (3.20)$$

The elastic coefficients are related to the force constants as shown in equation (2.30), with $c_{11}^{op} = c_{xxxx}$, $c_{12}^{op} = c_{xxyy}$, $c_{b44}^{op} = c_{xyyx}$, $c_{a44}^{op} = c_{xyxy}$. For the crystal to be isotropic, $c_{12}^{op} + c_{a44}^{op} + c_{b44}^{op} = c_{11}^{op}$. If we make this assumption,

$$\omega^2 \bar{\rho} u_x = \omega_T^2 \bar{\rho} u_x + \frac{\omega_T^2 \bar{\rho}(\varepsilon_s - \varepsilon_\infty)}{[\varepsilon_\infty - \varepsilon(\omega)]} u_x + [c_{11}^{op} \nabla.\mathbf{u}]_{,x} - [\nabla \times c_{a44}^{op} \nabla \times \mathbf{u}]_{,x} \qquad (3.21)$$

Furthermore, isotropy allows the connection with the acoustic case:

$$c_{11}^{op} = \frac{\mu}{M}c_{11}, \quad c_{a44}^{op} = \frac{\mu}{M}c_{44} \tag{3.22}$$

In terms of the longitudinal and transverse velocities, $c_{11} = \rho v_L^2$, $c_{44} = \rho v_T^2$; hence the velocities that determine dispersion are just those of acoustic modes (equation (2.35), etc.).

The inclusion of dispersion in equation (3.15) alters the dielectric function, which becomes different for longitudinal and transverse modes and varies with wave vector:

$$\varepsilon(\omega)_{longitudinal} = \varepsilon_\infty \frac{\omega^2 - \omega_L^2 + v_L^2 k^2}{\omega^2 - \omega_T^2 + v_L^2 k^2} \tag{3.23a}$$

$$\varepsilon(\omega)_{transverse} = \varepsilon_\infty \frac{\omega^2 - \omega_L^2 + v_T^2 k^2}{\omega^2 - \omega_T^2 + v_T^2 k^2}$$

Since the long-wavelength dispersion of the LO mode can be taken to be $\omega^2 = \omega_L^2 - v_L^2 k^2$, the permittivity remains zero, independent of wave vector.

Besides supporting LO and TO modes, the crystal, acting as a dielectric continuum, supports electric interface (IF) modes that have the dielectric function of transverse waves, but with $k = 0$. The dielectric function for the IF mode becomes simply

$$\varepsilon(\omega)_{interface} = \varepsilon_\infty \frac{\omega^2 - \omega_L^2}{\omega^2 - \omega_T^2} \tag{3.23b}$$

The field factors $s = \mathbf{E}/\mathbf{u}$ are

$$LO\, s = \frac{\omega^2 - \omega_T^2 + v_L^2 k^2}{\omega_L^2 - \omega_T^2} = 1$$

$$TO\, s = 0$$

$$IF\, s = \frac{\omega^2 - \omega_T^2}{\omega_L^2 - \omega_T^2} \tag{3.23c}$$

In the case of interface (IF) modes, $k^2 = k_x^2 + k_z^2 \approx 0$ (see Section 3.3).

3.3 Interface Modes

Maxwell's equations in a non-magnetic medium, in conventional notation, are

$$\nabla \times \mathbf{H} = \mathbf{j} + \frac{\partial \mathbf{D}}{\partial t}$$

$$\nabla \times \mathbf{E} = -\frac{\partial \mathbf{B}}{\partial t}$$

$$\nabla \cdot \mathbf{B} = 0$$

$$\nabla \cdot \mathbf{D} = 0 \tag{3.24}$$

At the comparably low frequencies of long-wavelength modes, the rates for **B** and **D** are negligible and so, to a good approximation, we can take $\nabla \times \mathbf{E} \approx 0$, which means that we can express the field in terms of a scalar potential:

$$\mathbf{E} = -\nabla \phi \tag{3.25}$$

The dispersion relation for a wave that travels with wave-vector components k_x, k_y, and k_z is

$$\omega^2 = \frac{k_x^2 + k_y^2 + k_z^2}{\varepsilon(\omega)\mu_0} = c^2 \mathbf{k}^2 \tag{3.26}$$

Here, c is the velocity of light in vacuo. For interface frequencies in the vicinity of the LO frequency, the relatively large magnitude of the velocity of light means that $k^2 \sim 0$. We have chosen k_x and k_y to be in the plane of the interface, and k_z to be perpendicular. In the case of electrical waves deriving from ionic polarization, the frequency remains near to the LO centre-zone value even when the wave vector increases, a condition achievable if k_z is imaginary and whose square is approximately equal to $k_x^2 + k_y^2$. The resultant is an electric wave that travels along the interface and is evanescent in the z direction. Taking the interface to be isotropic, we can chose k_x to be the vector along the surface and $k_z = ik_x$. The components of the interface wave, taking account of its transverse nature, are

$$\left.\begin{array}{l} u_x = Ak_x e^{-k_x z} e^{i(k_x x - \omega t)} \\ u_z = Aik_z e^{-k_x z} e^{i(k_x x - \omega t)} \end{array}\right\} \quad z > 0 \tag{3.27}$$

which has zero divergence and zero curl. A similar form, with appropriate change of signs, is applicable for $z < 0$.

3.4 Velocities

In an isotropic crystal, the velocities of optical phonons that determine the dispersion were shown to be equal to the acoustic velocities. In the real world this is not the case. In Section 2.4 of Chapter 2, we obtained the equivalent results for optical modes. The velocities are defined according to

$$v_{L,T} = \frac{\omega_{L,T} - \omega}{k} \tag{3.28}$$

Along the principal directions they are given by

$$\begin{array}{lll} <100> & <110> & <111> \\ \bar{\rho}v_L^2 = c_{11}^{op}, & \bar{\rho}v_L^2 = c_{11}^{op} - c_0/2 & \bar{\rho}v_L^2 = c_{11}^{op} - 2c_0/3 \\ \bar{\rho}v_T^2 = c_{44}^{op2} & \bar{\rho}v_T^2 = c_{44}^{op2} + c_0/2 & \bar{\rho}v_T^2 = c_{44}^{op2} + c_0/3 \\ \bar{\rho}v_T^2 = c_{44}^{op2} & \bar{\rho}v_T^2 = c_{44}^{op2} & \bar{\rho}v_T^2 = c_{44}^{op2} + c_0/3 \end{array} \tag{3.29}$$

In terms of the acoustic velocities $v_{L,T}$, we get

$$\begin{array}{ccc} <100> & <110> & <110> \\ v_L^2 - \beta'/4a\bar{\rho} & v_L^2 - \beta'/4a\bar{\rho} - c_0/2\bar{\rho} & v_L^2 - \beta'/4a\bar{\rho} - 2c_0/3\bar{\rho} \\ v_T^2 & v_T^2 + c_0/2\bar{\rho} & v_T^2 + c_0/3\bar{\rho} \\ v_T^2 & v_T^2 & v_T^2 + c_0/3\bar{\rho} \end{array} \tag{3.30}$$

Here, the isotropy coefficient is

$$c_0 = c_{11}^{op} - c_{12}^{op} - c_{44}^{op1} - c_{44}^{op2} = \frac{\mu}{aM}\left(\frac{\beta(\beta-\alpha)}{(\alpha+\beta)}\right)$$

$$\beta' = \beta - 2r\Delta\beta, \quad r = \frac{(m^A - m^B)}{2M}, \quad \Delta\beta = \frac{\beta^A - \beta^B}{2} \tag{3.31}$$

Isotropy implies that $\beta = 0$; this assumes that we can neglect all but nearest-neighbour interactions. It is evident that without the assumption of isotropy, life gets very hard.

3.5 Inhomogenous Material

The appropriate boundary conditions for optical modes at an abrupt interface has been a topic of some concern in nanostructure physics. Certain similarities between the properties of acoustic and optical modes have suggested that the same connection rules should apply, but the microscopic model of Section 2.6 in Chapter 2 indicates otherwise. There are significant differences in the respective equations of motion in the two cases: one is the duplication of the shear modulus, another is the involvement of mass factors in the moduli. Whereas, as we have argued, differences in the acoustic constants between lattice-matched materials may be ignored, differences in mass cannot be neglected. Using a one-dimensional model of the lattice, Akero and Ando (1989) were the first to point out that mass factors must certainly play a part in the connection rules for optical modes, and this has been amply confirmed by the microscopic model. Boundary conditions for optical modes are therefore bound to be more complex than for acoustic modes. In order to clarify the issues involved it is sufficient to focus solely on the non-polar factors in Chapter 4.

3.6 Piezoelectricity

The polarity of the acoustic modes is quantified by the piezoelectric constant tensor e_{ikl}, such that the electric displacement is

$$D_i = \varepsilon_{ij}E_j + e_{ikl}S_{kl} \tag{3.32}$$

(sums over repeated subscripts implied). Limiting our attention to the cubic zinc blende lattice reduces the permittivity tensor to ε_s, the static permittivity, and $e_{ijl} \to e_{14} = e_{15} = e_{16}$ in reduced notation. The shear components of the stress become

$$T_{xz} \equiv T_4 = c_{44}S_4 - e_{14}E_x$$
$$T_{zy} \equiv T_5 = c_{44}S_5 - e_{14}E_y$$
$$T_{yx} \equiv T_6 = c_{44}S_6 - e_{14}E_z \quad (3.33)$$

With $D = 0$, equation (3.32) yields the relation between field and strain:

$$E_x = -\frac{e_{14}}{\varepsilon_s}S_4, \text{etc.} \quad (3.34)$$

Substitution into equation (3.33) shows that the effect of piezoelectricity is to 'stiffen' the shear elastic constant:

$$c_{44} \to c_{44}^* = c_{44}(1 + K^2)$$
$$K^2 = \frac{e_{14}^2}{c_{44}\varepsilon_s} \quad (3.35)$$

Here, K is known as the electromechanical coefficient.

The equation of motion becomes

$$\rho \ddot{\mathbf{U}} = c_{44}^* \nabla^2 \mathbf{U} + (c_{12} + c_{44}^*)\nabla(\nabla.\mathbf{U}) + c_0^* \left(\hat{\mathbf{x}} \frac{\partial^2 U_x}{\partial x^2} + \hat{\mathbf{y}} \frac{\partial^2 U_y}{\partial y^2} + \hat{\mathbf{z}} \frac{\partial^2 U_z}{\partial z^2} \right) \quad (3.36)$$

$$c_0^* = c_{11} - c_{12} - 2c_{44}^*$$

The velocities along the principal crystallographic axes are obtained following the procedure of Section 1.3 in Chapter 1:

$$\begin{aligned}
[100] \quad & \rho v_L^2 = c_{12} + 2c_{44}^* + c_0^* = c_{11} \\
& \rho v_T^2 = c_{44}^* \\
& \rho v_T^2 = c_{44}^* \\
[110] \quad & \rho v_L^2 = c_{12} + 2c_{44}^* + c_0^*/2 \\
& \rho v_T^2 = c_{44}^* + c_0^*/2 \\
& \rho v_T^2 = c_{44}^* \\
[111] \quad & \rho v_L^2 = c_{12} + 2c_{44}^* + c_0^*/3 \\
& \rho v_T^2 = c_{44}^* + c_0^*/3 \\
& \rho v_T^2 = c_{44}^* + c_0^*/3
\end{aligned} \quad (3.37)$$

If elastic isotropy is assumed, it would seem to imply that $c_0^* = -2c_{44}K^2$. This would be wrong. Piezoelectricity is intimately associated with the intrinsic asymmetry of the lattice; in the case of the cubic sphalerite lattice of GaAs, etc., it is the lack of inversion symmetry that introduces an electric polarization in the presence of mechanical stress or electric field. The assumption of isotropy therefore implies that the crystal is non-piezoelectric and, therefore, $c_0^* = 0$.

The energy associated with the piezoelectricity is small compared with the mechanical energy, and it is often assumed that the non-polar mechanical boundary conditions are adequate. However, the piezoelectric interaction with electrons cannot be neglected at low temperatures. Under these circumstances, it is necessary to take into account elastic anisotropy in the description of acoustic modes in nanostructures.

4
Boundary Conditions

4.1 Introduction

An immediate problem that presented itself in the physics of nanostructures formed from two semiconductors was to describe conditions at their interface. In principle, band-structure and lattice-dynamical calculations could be performed to give the eigenvalues and eigenstates of the whole heterostructure. But, besides being computer-intensive, any result along these lines would refer only to the special case considered and would offer little in the way of providing criteria for predicting the properties of the huge number of structures that were conceivable. A less cumbersome approach, though inevitably more approximate, was to examine ways of connecting the well-established properties of the individual materials across the interface and to formulate connection rules.

The situation regarding acoustic modes had, of course, been a cornerstone of classical physics. Nevertheless, we will outline the steps that lead to the classical connections rules since it will be a good introduction to our discussion of optical modes, connection rules for which have been available only fairly recently.

4.2 Acoustic Modes

The equations of motion for non-polar acoustic modes neglecting all higher order differentials are, with the comma implying differentiation,

$$\begin{aligned}
-\omega^2 \rho U_x &= T_{xx,x} + T_{xy,y} + T_{xz,z} \\
-\omega^2 \rho U_y &= T_{yx,x} + T_{yy,y} + T_{yz,z} \\
-\omega^2 \rho U_z &= T_{zx,x} + T_{zy,y} + T_{zz,z}
\end{aligned} \quad (4.1)$$

Taking the interface to be normal to the z direction, we have to consider the connection rules for T_{xz}, T_{yz}, and T_{zz}. We can integrate over the interface, taking the interface to be roughly 2ε wide, and obtain

Hybrid Phonons in Nanostructures. First Edition. B.K. Ridley. © B.K. Ridley 2017.
Published in 2017 by Oxford University Press. DOI: 10.1093/acprof:oso/9780198788362.001.0001

$$T_{xz}(+\varepsilon) = T_{xz}(-\varepsilon) - \int_{-\varepsilon}^{\varepsilon} (\omega^2 \rho U_x + T_{xx,x} + T_{xy,y}) dz$$

$$T_{yz}(+\varepsilon) = T_{yz}(-\varepsilon) - \int_{-\varepsilon}^{\varepsilon} (\omega^2 \rho U_y + T_{yx,x} + T_{yy,y}) dz \quad (4.2)$$

$$T_{zz}(+\varepsilon) = T_{zz}(-\varepsilon) - \int_{-\varepsilon}^{\varepsilon} (\omega^2 \rho U_z + T_{zx,x} + T_{zy,y}) dz$$

In quasi-continuum theory, discrete entities are rigorously converted to continuous functions with Fourier components restricted to the first Brillouin zone. The stress factors in the integrand are all of order k^2, and therefore negligible in the long-wavelength approximation. Taking $\varepsilon \to 0$ makes the integrals vanish and we obtain the classical connection rule for the stress factors, that is, the continuity of

$$T_{xz}, T_{yz}, T_{zz} \quad (4.3)$$

In terms of the spatial displacements, these are

$$\begin{aligned} T_{xz} &= c_{44}(U_{z,x} + U_{x,z}) \\ T_{yz} &= c_{44}(U_{y,z} + U_{z,y}) \\ T_{zz} &= c_{11}U_{z,z} + c_{12}(U_{x,x} + U_{y,y}) \end{aligned} \quad (4.4)$$

The connection rules for the displacement components are derived from the integration:

$$U_x(+\varepsilon) = U_x(-\varepsilon) + \int_{-\varepsilon}^{\varepsilon} (c_{44})^{-1}(T_{xz} - U_{z,x}) dz$$

$$U_y(+\varepsilon) = U_y(-\varepsilon) + \int_{-\varepsilon}^{\varepsilon} (c_{44})^{-1}(T_{yz} - U_{z,y}) dz \quad (4.5)$$

$$U_z(+\varepsilon) = U_z(-\varepsilon) + \int_{-\varepsilon}^{\varepsilon} (c_{11})^{-1}(T_{zz} - c_{12}(U_{x,x} + U_{y,y})) dz$$

In the long-wavelength approximation, we recover the classical connection rule for amplitude, that is, the continuity of

$$U \quad (4.6)$$

4.3 Optical Modes

As mentioned in Section 2.6, if we consider a lattice-matched structure such as AlAs/GaAs, in which the interface is defined by the atom common to both semiconductors, namely As, then it would seem that the acoustic boundary conditions should apply. The displacement **u** of the arsenic atom is what it is, independent of whether it is GaAs or AlAs. In this sense the continuity is guaranteed. Equally, the atom does not move out of the interface in response to a force, which implies that the net force acting on the atom is zero. Since a net force is associated with a variation of stress, it means that the stress must be continuous across the interface. The connection rules are therefore the acoustic ones.

What is wrong with this argument is the assumption that the interface can be defined by the position of a particular atom. In the case of AlAs/GaAs, the arsenic atom has bonds with gallium atoms on one side and aluminium atoms on the other. The physical nature of the interface is therefore more diffuse, involving the variation of bond strengths and mass. In the acoustic case, the elastic constants vary little from one semiconductor to the other, and mass is not involved.

The situation for optical modes was very different in that there was no classical result available. Early guesses at the connection rules were to identify them with the acoustic ones (Perez-Alvarez et al. 1993; Ridley et al. 1994), that is, the continuity of

$$\mathbf{u}, T^o_{xz}, T^o_{yz}, T^o_{zz} \tag{4.7}$$

However, this guess is not a good one unless the force constant and the optical elastic constants are very slowly varying across the interface. As mentioned previously in connection with the acoustic modes (Section 1.5), it is not unreasonable to assume that the force constants do not vary appreciably from semiconductor to semiconductor, but in the optical case, elastic constants involve the mass, so these functions are very different and their variation across an interface introduces delta function terms that make the acoustic analogue utterly invalid. Moreover, the disparity in frequency, absent in the acoustic case, is often too large for an optical wave in one semiconductor to travel further than the first atomic layer in another. In this case, the connection rule reduces effectively to

$$u = 0 \tag{4.8}$$

This rule is obviously also appropriate at a free surface.

In our subsequent discussion in later chapters of hybrid modes in nanostructures, we will assume that this connection rule holds good for the materials involved. However, there are nanostructures involving alloys, for example, $Ga_{1-x}Al_xAs/GaAs$, and in others involving similar frequencies, for example, InAs/GaSb, where the simple connection rule of equation (4.11) is not correct. More general connection rules have been explored by Akero and Ando (1989), who were the first to point out the role of mass, and by Foreman (1995, 1998). We give a brief account of the latter's approach.

Once again we consider the interface to be normal to the z direction, of finite width, and to consist of adjacent non-polar semiconductors. (The addition of a polar component to the restoring force can readily be taken into account, but it cannot affect substantially the results for the mechanical boundary conditions: the electrical energy density is much smaller than the mechanical energy density in polar semiconductors.)

The equation of motion for optical vibrations is

$$\omega^2 \bar{\rho} u_\alpha = c^o_{\alpha\beta} u_\beta + (c^0_{\alpha\lambda\beta\mu} u_{\beta,\mu})_{,\lambda} \tag{4.9}$$

Here $\bar{\rho}$ is the reduced density and, as in Chapter 2, the comma denotes differentiation and repeated suffixes imply addition. The force constant for inhomogeneous material has been given in Section 2.6 of Chapter 2, repeated here for convenience:

$$c_{zz} = \frac{\alpha+\beta}{4a^3} + \frac{\alpha+3\beta}{8a}\left[\left(\frac{\mu}{M}\right)_{,zz} + 2(r_{,z})^2\right] - \frac{\Delta\beta}{4a}r_{,zz}$$

$$c_{xx} = c_{yy} = \frac{\alpha+\beta}{4a^3} + \frac{\alpha+\beta}{8a}\left[\left(\frac{\mu}{M}\right)_{,zz} + 2(r_{,z})^2\right] \tag{4.10}$$

$$c_{xy} = c_{yx} = \frac{\alpha-\beta}{4a^2}r_{,z}$$

The optical elastic constants in homogeneous material are

$$c_{xx} = \frac{\alpha+\beta}{4a^3},$$

$$c^{op}_{11} = \frac{\mu}{M}\frac{\alpha+3\beta}{4a} - \frac{\beta'}{4a}$$

$$c^{op}_{12} = \frac{\mu}{M}\frac{\alpha}{4a} - \frac{\beta'}{8a}$$

$$c^{op}_{a44} = \frac{\mu}{M}\frac{\alpha+\beta}{4a} - \frac{(\alpha-\beta)^2}{a\omega^2 M} \tag{4.11}$$

$$c^{op}_{b44} = \frac{\mu}{M}\frac{\alpha}{4a} - \frac{\beta'}{8a} - \frac{(\alpha-\beta)^2}{a\omega^2 M}$$

The components of stress acting across the interface are

$$T_{xz} = c^o_{a44} u_{x,z} + c^o_{b44} u_{z,x}$$

$$T_{yz} = c^o_{a44} u_{y,z} + c^o_{b44} u_{z,y} \tag{4.12}$$

$$T_{zz} = c^o_{11} u_{z,z} + c^o_{12}(u_{x,x} + u_{y,y})$$

The equations of motion are, explicitly,

$$\omega^2 \bar{\rho} u_x = c_{xx} u_x + c_{xy} u_y + T_{xx,x} + T_{xy,y} + T_{xz,z}$$
$$\omega^2 \bar{\rho} u_y = c_{yy} u_y + c_{yx} u_x + T_{yy,y} + T_{yz,z} + T_{yx,x} \quad (4.13)$$
$$\omega^2 \bar{\rho} u_z = c_{zz} u_z + T_{zz,z} + T_{zx,x} + T_{zy,y}$$

Integrating the equation of motion gives

$$T_{xz}(+\varepsilon) = T_{xz}(-\varepsilon) - \int_{-\varepsilon}^{\varepsilon} (Q_{xx} u_x + Q_{xy} u_y + T_{xx,x} + T_{xy,y}) dz$$

$$T_{yz}(+\varepsilon) = T_{yz}(-\varepsilon) - \int_{-\varepsilon}^{\varepsilon} (Q_{yy} u_y + Q_{yx} u_x + T_{yy,y} + T_{yx,x}) dz \quad (4.14)$$

$$T_{zz}(+\varepsilon) = T_{zz}(-\varepsilon) - \int_{-\varepsilon}^{\varepsilon} (Q_{zz} u_z + T_{zx,x} + T_{zy,y}) dz$$

$$Q_{\alpha\alpha} = c_{\alpha\alpha} - \omega^2 \bar{\rho}, \quad Q_{xy} = c_{xy} \quad (4.15)$$

For long wavelengths, the stress terms in the integrands will be much smaller than the Q factors, which tend to delta functions as $\varepsilon \to 0$.

The displacement is related to the stress, for example:

$$u_{x,z} = (c_{a44}^o)^{-1}(T_{xz} - c_{b44}^o u_{z,x}) \quad (4.16)$$

Integrating gives

$$u_x(+\varepsilon) = u_x(-\varepsilon) + \int_{-\varepsilon}^{\varepsilon} p_{44}(T_{xz} - c_{b44}^o u_{z,x}) dz$$

$$u_y(+\varepsilon) = u_y(-\varepsilon) + \int_{-\varepsilon}^{\varepsilon} p_{44}(T_{yz} - c_{b44}^o u_{z,y}) dz \quad (4.17)$$

$$u_z(+\varepsilon) = u_z(-\varepsilon) + \int_{-\varepsilon}^{\varepsilon} p_{11}[T_{zz} - c_{12}^o(u_{x,x} + u_{y,y})] dz$$

$$p_{44} = (c_{a44}^o)^{-1}, \quad p_{11} = (c_{11}^o)^{-1} \quad (4.18)$$

The variation of the displacements in the integrand can be neglected for long wavelengths.

Regarding the interface components as small, we can use perturbation theory to provide the connection rules. Partial integration of the terms involving p, and substituting the stress terms from the equations of motion (equation (4.13)) gives, to lowest order,

$$u_x(+\varepsilon) = u_x(-\varepsilon) - p_{44}(\varepsilon,-\varepsilon)T_{xz}(-\varepsilon) + \int_{-\varepsilon}^{\varepsilon} p_{44}(\varepsilon,z)(Q_{xx}u_x + Q_{xy}u_y)dz$$

$$u_y(+\varepsilon) = u_y(-\varepsilon) - p_{44}(\varepsilon,-\varepsilon)T_{yz}(-\varepsilon) + \int_{-\varepsilon}^{\varepsilon} p_{44}(\varepsilon,z)(Q_{yy}u_y + Q_{yx}u_x)dz \quad (4.19)$$

$$u_z(+\varepsilon) = u_z(-\varepsilon) - p_{11}(\varepsilon,-\varepsilon)T_{zz}(-\varepsilon) + \int_{-\varepsilon}^{\varepsilon} p_{11}(\varepsilon,z)Q_{zz}u_z dz$$

$$p_{44}(\varepsilon,z) = \int_z^{\varepsilon} p_{44} dz, \quad p_{11}(\varepsilon,z) = \int_z^{\varepsilon} p_{11} dz \quad (4.20)$$

Here, u_x and u_y can be substituted into equation (4.14) to get the connection rule for the stress. This procedure can be repeated to provide greater accuracy.

The connection rules are expressed succinctly in matrix form

$$\begin{pmatrix} u_\alpha \\ T_{\alpha z} \end{pmatrix}_{+\varepsilon} = \mathbf{T} \begin{pmatrix} u_\alpha \\ T_{\alpha z} \end{pmatrix}_{-\varepsilon} \quad (4.21)$$

To lowest order

$$\mathbf{T} = \begin{Bmatrix} 1+a_{xx} & a_{xy} & 0 & -d_{xx} & d_{xy} & 0 \\ a_{yx} & 1+a_{yy} & 0 & d_{yx} & -d_{yy} & 0 \\ 0 & 0 & 1+a_{zz} & 0 & 0 & -d_{zz} \\ -b_{xx} & -b_{xy} & 0 & 1-a_{xx} & -a_{xy} & 0 \\ -b_{yx} & -b_{yy} & 0 & -a_{yx} & 1-a_{yy} & 0 \\ 0 & 0 & -b_{zz} & 0 & 0 & 1-a_{zz} \end{Bmatrix} \quad (4.22)$$

$$a_{xx} = \int_{-\varepsilon}^{\varepsilon} p_{44}(\varepsilon,z)Q_{xx}dz \quad b_{xx} = \int_{-\varepsilon}^{\varepsilon} Q_{xx}dz \quad d_{xx} = p_{44}(\varepsilon,-\varepsilon) - \int_{-\varepsilon}^{\varepsilon}[p_{44}(\varepsilon,z)]^2 Q_{xx}dz$$

$$a_{xy} = \int_{-\varepsilon}^{\varepsilon} p_{44}(\varepsilon,z)Q_{xy}dz \quad b_{xy} = \int_{-\varepsilon}^{\varepsilon} Q_{xy}dz \quad d_{xy} = \int_{-\varepsilon}^{\varepsilon}[p_{44}(\varepsilon,z)]^2 Q_{xy}dz$$

$$a_{zz} = \int_{-\varepsilon}^{\varepsilon} p_{11}(\varepsilon,z)Q_{zz}dz \quad b_{zz} = \int_{-\varepsilon}^{\varepsilon} Q_{zz}dz \quad d_{zz} = p_{11}(\varepsilon,-\varepsilon) - \int_{-\varepsilon}^{\varepsilon}[p_{11}(\varepsilon,z)]^2 Q_{zz}dz$$

$$(4.23)$$

$$a_{xx} = a_{yy}, a_{xy} = a_{yx}, b_{xx} = b_{yy}, b_{xy} = b_{yx}, c_{xx} = c_{yy}, c_{xy} = c_{yx} \quad (4.24)$$

Given that the force factors are the same in adjacent material, the Q factors vary with the mass factors μ/M and r (equation (2.38)). In the spirit of the envelope-function method, we may describe the variation of the relevant factors by an error function:

$$f(z) = \bar{f} + \Delta f \mathrm{erf}(z/\sigma),$$

$$\bar{f} = \frac{1}{2}[f(+\varepsilon) + f(-\varepsilon)], \quad \Delta f = \frac{1}{2}[f(+\varepsilon) - f(-\varepsilon)] \quad (4.25)$$

Here, σ is a Gaussian half-width of the order of the dimension of the unit cell. The integrals can then be evaluated. To lowest order approximation, the results are

$$a_{xx} = -\frac{\alpha + \beta}{8a} \left\langle (c_{a44}^0)^{-1} \right\rangle \Delta \left(\frac{\mu}{M}\right)$$

$$a_{xy} = 0$$

$$a_{zz} = -\left\langle (c_{a44}^0)^{-1} \right\rangle \left(\frac{\alpha + 3\beta}{8a} \Delta \left(\frac{\mu}{M}\right) + \frac{\Delta \beta}{4a} \Delta r \right) \quad (4.26a)$$

$$b_{xx} = \frac{\alpha + \beta}{8a} \frac{\sqrt{(2/\pi)}}{\sigma} (\Delta r)^2$$

$$b_{xy} = \frac{\alpha - \beta}{4a^2} \Delta r$$

$$b_{zz} = \frac{\alpha + 3\beta}{8a} \frac{\sqrt{(2/\pi)}}{\sigma} (\Delta r)^2 \quad (4.26b)$$

$$d_{xx} = d_{xy} = d_{zz} = 0 \quad (4.26c)$$

The bracket <> denotes an average, and the delta denotes the difference in the mass terms across the interface. ($\Delta \beta$ is the difference of bond-bending force constants in the A and B superlattices (Table 2.2).)

Discontinuities in particle displacement can be seen to be driven by the difference in optical elastic constant associated with the difference in the factor μ/M. On the other hand, the discontinuity in stress is driven by the difference in the difference of ionic mass, as depicted by the $b_{\alpha\beta}$ coefficients, which are simply interface stresses.

The case of Group IV semiconductors is particularly simple: $\mu/M = 1/4, r = 0$, thus all the coefficients vanish and the boundary conditions are just those of the acoustic modes. The AlAs/GaAs system posseses a common anion, As, that serves to define the interface. If, instead of assuming an envelope-function approach, we adopt an abrupt-step model, in which the relevant variables are assumed to take in their average values at the interface, then letting the width of the interface region to approach zero allows the boundary conditions to be the classical ones for acoustic modes. However, the disparity in frequency must be taken into account. When the disparity is small, as it is in the case of $Al_xGa_{1-x}As$ with x small, the acoustic connection rules imply short wavelengths in the alloy which, if too short, would invalidate the fundamental long-wavelength assumption. The situation is similar for the AlN/GaN system, but without the restriction on x, there being substantial overlap of the LO bands (Bennett et al. 1999). Nevertheless, a large disparity in wavelength acts in the same way as a large disparity in frequency, both pointing towards the simple boundary condition: $\mathbf{u} = 0$ at the interface.

The case of InAs/GaSb presents a different problem, since there is no shared ion. The interface can now be a relatively extended region of pure GaAs, bounded on one side by an interface defined by the As anion and on the other by the Ga cation; or a region of pure InSb, bounded on one side by the In cation and on the other by the Sb anion. The system allows the existence of localized interface modes with their own characteristic frequency (Foreman 1995).

Assuming isotropy allows the connection matrix (equation 4.22) to be factored. With TO and LO modes now distinct, we can classify the modes that are incident on the interface into s-modes, having no component of displacement normal to the interface, and p-modes, having a component normal to the surface (Fig. 4.1).

Only TO modes can be s-modes, but both TO and LO modes can be p-modes. For an s-mode whose displacement is u_x, the matrix becomes

$$\begin{pmatrix} 1 + a_{xx} & -d_{xx} \\ -b_{xx} & 1 - a_{xx} \end{pmatrix} \qquad (4.27)$$

For a p-mode having components u_x and u_z:

$$\begin{pmatrix} 1 + a_{xx} & 0 & -d_{xx} & 0 \\ 0 & 1 + a_{zz} & 0 & -d_{zz} \\ -b_{xx} & 0 & 1 - a_{xx} & 0 \\ 0 & -b_{zz} & 0 & 1 - a_{zz} \end{pmatrix} \qquad (4.28)$$

Provided that the interface introduces a small perturbation, these results, which are to lowest order, will be sufficient.

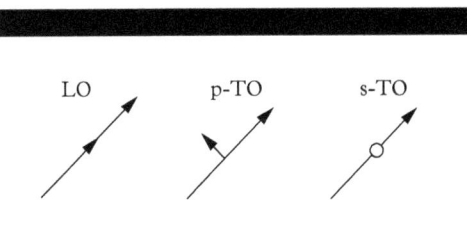

Figure 4.1 *s- and p-modes.*

4.4 Electromagnetic Boundary Conditions

The mechanical energy of lattice waves being so much greater than the electromagnetic energy in polar material, it is usual to apply mechanical boundary conditions as if the material were non-polar. However, the classical electromagnetic boundary conditions cannot be ignored. These are four in number and they require the continuity of:

(1) The tangential electric field **E** (or, equivalently, the scalar potential).
(2) The normal component of the electric displacement **D**.
(3) The tangential magnetic field **H**.
(4) The normal component of the magnetic induction **B**.

Given the relatively low frequencies of the lattice modes, the boundary conditions for the magnetic vectors are usually ignored in a quasi-static approximation.

5
Scalar and Vector Fields

5.1 Introduction

Those quantum structures describable in terms of Cartesian coordinates (e.g. the quantum well) can be treated with the minimum of formality. Turning to quantum structures describable in terms of curvilinear coordinates, we find it convenient to establish a more formal basis in order to describe the LO and TO modes in terms of scalar and vector potentials.

5.2 The Helmholtz Equation

The equation of motion in the isotropic approximation relating long-wavelength optical displacement to frequency is

$$v_L^2 \nabla(\nabla \cdot \mathbf{u}) - v_T^2 \nabla \times \nabla \times \mathbf{u} = -(\omega_{L,T}^2 - \omega^2)\mathbf{u} \qquad (5.1)$$

where $v_{L,T}$ are the velocities that determine the dispersion of LO and TO bulk modes, and $\omega_{L,T}$ are the zone-centre frequencies of the LO and TO modes. At this stage, it is usual to define $\mathbf{w} = \rho^{1/2}\mathbf{u}$, where ρ is the reduced-mass density, and relate $\mathbf{w}$ to the potentials. Here, we express the displacement in terms of a scalar potential and a vector potential:

$$\mathbf{u} = \nabla\phi + \nabla \times \mathbf{A} \qquad (5.2)$$

Substitution into equation (5.1) gives

$$v_L^2 \nabla \nabla^2 \phi - v_T^2 \nabla \times \nabla \times \nabla \times \mathbf{A} = -(\omega_{L,T}^2 - \omega^2)(\nabla\phi + \nabla \times \mathbf{A}) \qquad (5.3)$$

We have used the identities $\nabla \cdot (\nabla \times \mathbf{A}) = 0$, $\nabla \times (\nabla\phi) = 0$. Another identity is $\nabla \times \nabla \times \mathbf{A} = \nabla(\nabla \cdot \mathbf{A}) - \nabla^2 \mathbf{A}$. We choose the vector potential to satisfy $\nabla \cdot \mathbf{A} = 0$, and equation (5.3) divides into separate equations for longitudinal and transverse modes. Taking the vector potential to have zero divergence, we arrive at separate Helmholtz equations for the potentials:

$$\nabla^2 \phi = -\left(\frac{\omega_L^2 - \omega^2}{v_L^2}\right)\phi = -k_{LO}^2 \phi \tag{5.4a}$$

$$\nabla^2 \mathbf{A} = -\left(\frac{\omega_T^2 - \omega^2}{v_T^2}\right)\mathbf{A} = -k_{TO}^2 \mathbf{A} \tag{5.4b}$$

where $k_{LO,TO}$ are the wave vectors that determine the bulk dispersion of long waves. The scalar equation presents no problem, but the vector equation is more difficult in general, having analytic solutions only in a few cases (Morse and Feshbach 1953). Among these cases are the quantum well, etc., which are simply described in terms of Cartesian coordinates, the quantum wire in the form of a cylinder with circular cross section, and the quantum dot in the form of a sphere.

Here we consider the problem of relating the displacement of the transversely polarized modes to the vector potential in the case of the cylinder and of the sphere.

5.3 Cylinder

The equation for the vector potential is (equation (5.4b)) as follows:

$$\nabla^2 \mathbf{A} = -k_{TO}^2 \mathbf{A} \tag{5.5}$$

In terms of the relevant unit vectors

$$\mathbf{A} = A_r \hat{\mathbf{r}} + A_\theta \hat{\boldsymbol{\theta}} + A_z \hat{\mathbf{z}} \tag{5.6}$$

Unlike the case with rectangular coordinates, the unit vectors can vary with the coordinate position in the case of curvilinear coordinates. This makes the treatment of the vector potential more complicated than it was for the scalar potential. In the cylindrical coordinate system, the unit vectors do not vary with changes in the radial or axial components, but for changes in the angular component we have

$$\frac{\partial \hat{\mathbf{r}}}{\partial \theta} = \hat{\boldsymbol{\theta}}, \quad \frac{\partial \hat{\boldsymbol{\theta}}}{\partial \theta} = -\hat{\mathbf{r}}, \quad \frac{\partial \hat{\mathbf{z}}}{\partial \theta} = 0 \tag{5.7}$$

Equation (5.5) becomes

$$\left[\nabla^2 A_r - \frac{2}{r^2}\frac{\partial A_\theta}{\partial \theta} - \frac{A_r}{r^2}\right]\hat{\mathbf{r}} + \left[\nabla^2 A_\theta + \frac{2}{r^2}\frac{\partial A_r}{\partial \theta} - \frac{A_\theta}{r^2}\right]\hat{\boldsymbol{\theta}} + \nabla^2 A_z = -k_{TO}^2 \mathbf{A} \tag{5.8}$$

This is not a pleasant equation. However, a solution may be found by considering the vector

$$\mathbf{A} = \psi \hat{\mathbf{z}}$$

where ψ is a scalar quantity and a solution of the equation for the scalar field:

$$\nabla^2\psi = -k_{TO}^2\psi \tag{5.9}$$

However, this vector will not serve since it does not have zero divergence, but we can form two independent solutions that do satisfy the condition of vanishing divergence, namely:

$$\begin{aligned}\mathbf{A}_1 &= \nabla \times \psi\hat{\mathbf{z}} \\ \mathbf{A}_2 &= \frac{1}{k}\nabla \times \mathbf{A}_1\end{aligned} \tag{5.10}$$

The vectors of equation (5.10) are indeed solutions of equation (5.8). From these we can get the particle displacements for the two orthogonal TO modes (Chapter 9).

5.4 Sphere

In the case of a sphere, the unit vectors all become functions of the angular coordinates:

$$\begin{aligned}\frac{\partial\hat{\mathbf{r}}}{\partial\theta} &= \hat{\theta}, \quad \frac{\partial\hat{\theta}}{\partial\theta} = -\hat{\mathbf{r}}, \quad \frac{\partial\hat{\phi}}{\partial\theta} = 0 \\ \frac{\partial\hat{\mathbf{r}}}{\partial\phi} &= \hat{\phi}\sin\theta, \quad \frac{\partial\hat{\theta}}{\partial\phi} = \hat{\phi}\cos\theta, \quad \frac{\partial\hat{\phi}}{\partial\phi} = -\hat{\mathbf{r}}\sin\theta - \hat{\theta}\cos\theta\end{aligned} \tag{5.11}$$

From equation (5.4b) we obtain, taking account of the fact that unit vectors vary with angles as above:

$$\begin{aligned}(\nabla^2\mathbf{A})_{\mathbf{r}} &= \nabla^2 A_r - \frac{2}{r^2}A_r - \frac{2}{r^2}\frac{\partial A_\theta}{\partial\theta} - \frac{2\cos\theta}{r^2\sin\theta}A_\theta - \frac{2}{r^2\sin\theta}\frac{\partial A_\varphi}{\partial\varphi} = k^2 A_r \\ (\nabla^2\mathbf{A})_\theta &= \nabla^2 A_\theta - \frac{1}{r^2\sin^2\theta}A_\theta + \frac{2}{r^2}\frac{\partial A_r}{\partial\theta} - \frac{2\cos\theta}{r^2\sin^2\theta}\frac{\partial A_\varphi}{\partial\varphi} = k^2 A_\theta \\ (\nabla^2\mathbf{A})_\varphi &= \nabla^2 A_\varphi - \frac{1}{r^2\sin^2\theta}A_\varphi + \frac{2}{r^2\sin\theta}\frac{\partial A_r}{\partial\varphi} + \frac{2\cos\theta}{r^2\sin^2\theta}\frac{\partial A_\theta}{\partial\varphi} = k^2 A_\varphi\end{aligned} \tag{5.12}$$

This is awfully messy, but, as in the case of the cylinder, solutions can be found by considering the vector:

$$\mathbf{A} = r\psi\hat{\mathbf{r}} \tag{5.13}$$

Once again, ψ is the solution of the corresponding scalar equation. The two solutions, each with zero divergence, are

$$\mathbf{A}_1 = \nabla \times (r\psi \hat{\mathbf{r}})$$
$$\mathbf{A}_2 = \frac{1}{k} \nabla \times \mathbf{A}_1 \tag{5.14}$$

from which the particle displacements of the two TO modes can be deduced (Chapter 10).

Part 2

Hybrid Modes in Nanostructures

6
Non-Polar Slab

6.1 Boundary Conditions

Hybrid modes exist as a consequence of acoustic and optical waves having to satisfy the boundary conditions at an interface or at a surface. We adopt several simplifying assumptions, namely:

(1) The long-wavelength approximation—decouples acoustic and optical modes.
(2) The crystals are isotropic—decouples LO and TO modes.
(3) The disparity of the optical mode frequencies across the interface is large, so that the appropriate boundary condition that achieves mechanical equilibrium is $\mathbf{u} = 0$.

We begin the description of hybrid modes in nanostructures with an account of modes in a non-polar, free-standing slab. (The case of polar modes is discussed in the context of a quantum well in Chapter 8.)

We consider the slab to be bounded in the z direction at $z = \pm L/2$, and unbounded otherwise. Because of our assumption of elastic isotropy, we can distinguish between longitudinally polarized modes, abbreviated by L (LA for acoustic modes and LO for optical modes), and transversely polarized modes, abbreviated by T (TA and TO). It is convenient to distinguish transverse modes with displacement in the plane of incidence—pT—and transverse modes with polarization perpendicular to the plane of incidence—sT (see Fig. 6.1). The surface has coordinates x and y and the normal to the plane has the coordinate z. Having assumed isotropy, we note that there is no further lack of generality in adopting this coordinate scheme.

The free surface allows any sort of acoustic displacement and any sort of stress within the surface, but not any stress that exerts forces perpendicular to the surface, since these cannot be matched in the surrounding vacuum. Of the dilatational stresses, T_3, and of the shear stresses, T_4 and T_5, are the ones that have this property; the boundary conditions are, therefore,

6 Non-Polar Slab

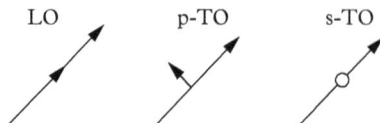

Figure 6.1 *s*- and *p*-modes.

$$T_3 = c_{11}\frac{\partial u_z}{\partial z} + c_{12}\left(\frac{\partial u_x}{\partial x} + \frac{\partial u_y}{\partial y}\right) = c_{11}\nabla \cdot \mathbf{u} - 2c_{44}\left(\frac{\partial u_x}{\partial x} + \frac{\partial u_y}{\partial y}\right) = 0$$
$$T_4 = c_{44}\left(\frac{\partial u_y}{\partial z} + \frac{\partial u_z}{\partial y}\right) = 0 \qquad (6.1)$$
$$T_5 = c_{44}\left(\frac{\partial u_z}{\partial x} + \frac{\partial u_x}{\partial z}\right) = 0$$

The sign of these equivalent stresses is negative for optical modes, but that is irrelevant here since it is the force between the vibrating atoms that is most important. For optical modes the surface atoms do not share the crystal symmetry of the bulk nor the same force constants. The transition of an optical wave from bulk to surface is more likely to be similar to encountering a very rigid material than it is for an acoustic wave, which suggests that the boundary condition in this case is $\mathbf{u} = 0$. We will assume that this is the case for most systems.

6.2 Acoustic Modes

6.2.1 s Modes (Love Waves)

The displacement that describes the sTA modes (Love 1920) has the form

$$u_y = e^{ik_x x}\left(Ae^{ik_z z} \pm Be^{-ik_z z}\right) \qquad (6.2)$$

Its divergence is zero and it automatically satisfies the boundary conditions for T_3 and T_5. The boundary condition for T_4 is satisfied by

$$u_y = Ae^{ik_x x}\begin{cases}\sin k_z z & k_z L = (2n+1)\pi \\ \cos k_z z & k_z L = 2n\pi\end{cases} \qquad (6.3)$$

with n an integer. In general, energy normalization gives

$$A = \frac{1}{K}\left(\frac{\hbar}{2M\omega}\right)^{1/2}(a + a^\dagger) \qquad (6.4)$$

with K determined by an integral over the mechanical energy in the cavity. In the case of the sTA mode,

$$K^2 = \frac{1}{2} \tag{6.5}$$

The dispersion relation for the s mode is $\omega^2 = v_T^2(k_x^2 + k_z^2)$. For the symmetric case, the smallest value of k_z is π/L. Thus, for $\omega < v_T\pi/L$, k_x becomes imaginary and the wave becomes non-propagating.

6.2.2 Lamb Waves

Neither of the p modes (Lamb 1910) can satisfy the boundary conditions on its own, but satisfaction is possible when combined in a phase-matched double hybrid. Such hybrids are known as Lamb waves. We label the modes symmetric or antisymmetric with respect to the x component. Thus, for the antisymmetric mode:

$$\begin{aligned} u_x &= e^{ik_x x}(k_x A \sin k_L z + k_T B \sin k_T z) \\ u_z &= ie^{ik_x x}(-k_L A \cos k_L z + k_x B \cos k_T z) \end{aligned} \tag{6.6}$$

Here, A is the amplitude for the pLA mode, and B for the pTA mode. The z components of the wave vectors are labelled k_L and k_T. The condition for T_4 is satisfied automatically. The boundary conditions for T_3 and T_5 at $z = \pm L/2$ give two simultaneous equations for the amplitudes, which have solutions provided the determinant of the coefficients vanishes, which gives

$$\begin{aligned} &-pt\sin(k_L L/2)\cos(k_T L/2) + qr\cos(k_L L/2)\sin(k_T L/2) = 0 \\ &\text{i.e. } \tan(k_L L/2) = (qr/pt)\tan(k_T L/2) \end{aligned} \tag{6.7}$$

$$\begin{aligned} p &= (1-2\gamma)k_x^2 + k_L^2, \ q = 2\gamma k_x k_T, \ r = 2k_x k_L, \ t = k_x^2 - k_L^2 \\ \gamma &= c_{44}/c_{11} \end{aligned} \tag{6.8}$$

The symmetric solution is

$$\begin{aligned} u_x &= e^{ik_x x}(k_x A \cos k_L z + k_T B \cos k_T z) \\ u_z &= ie^{ik_x x}(k_L A \sin k_L z - k_x B \sin k_T z) \end{aligned} \tag{6.9}$$

The corresponding determinant for the stresses gives

$$\begin{aligned} &-qr\sin(k_L L/2)\cos(k_T L/2) + pt\cos(k_L L/2)\sin(k_T L/2) = 0 \\ &\text{i.e. } \cot(k_L L/2) = (qr/pt)\cot(k_T L/2) \end{aligned} \tag{6.10}$$

There is one further condition to be fulfilled; namely, that both components of the pLA–TA hybrid have the same frequency. This means that

$$\omega^2 = v_L^2(k_x^2 + k_L^2) = v_T^2(k_x^2 + k_T^2) \tag{6.11}$$

In both cases, the condition restricting the z components of the wave vectors imposes a restriction on the x component.

It is useful to define a slab velocity $\mathbf{v}$ and an associated parameter s as follows:

$$\begin{aligned}\omega^2 &= v^2 k_x^2 \\ s &= v^2/v_T^2 \end{aligned} \tag{6.12}$$

Thus

$$k_L^2 = (\gamma s - 1)k_x^2, \quad k_T^2 = (s-1)k_x^2 \tag{6.13}$$

where we have used $\gamma = v_T^2/v_L^2$. In terms of the slab velocity, the parameters defined in equation (6.8) are (suppressing the in-plane wave vector) as follows:

$$p = \gamma(s-2), \quad q = 2\gamma\sqrt{s-1}, \quad r = 2\sqrt{\gamma s - 1}, \quad t = 2-s \tag{6.14}$$

Equations (6.7) and (6.10), along with the restriction equation (6.11), define a set of guided modes and a set of surface modes. In general, these equations must be solved numerically.

6.2.2.1 Guided Modes

The solutions with real wave vectors that satisfy the stress conditions are guided modes, consisting of an antisymmetric mode (symmetric being defined by the pattern of the z direction of the x component):

$$\begin{aligned} u_x &= Ae^{ik_x x}\left(k_x \sin k_L z + k_T \frac{r\cos(k_L L/2)}{t\cos(k_T L/2)} \sin k_T z\right) \\ u_z &= iAe^{ik_x x}\left(-k_L \cos k_L z + k_x \frac{r\cos(k_L L/2)}{t\cos(k_T L/2)} \cos k_T z\right) \end{aligned} \tag{6.15}$$

and a symmetric mode:

$$\begin{aligned} u_x &= Ae^{ik_x x}\left(k_x \cos k_L z + k_T \frac{r\sin(k_L L/2)}{t\sin(k_T L/2)} \cos k_T z\right) \\ u_z &= iAe^{ik_x x}\left(k_L \sin k_L z - k_x \frac{r\sin(k_L L/2)}{t\sin(k_T L/2)} \sin k_T z\right) \end{aligned} \tag{6.16}$$

For small k_x, equations (6.7) and (6.10) are satisfied by

$$\left.\begin{array}{l} LA/pTA\ k_L L = n\pi,\ n\ \text{even} \\ pTA/LA\ k_T L = n\pi,\ n\ \text{odd} \end{array}\right\} \text{eqn (6.7)}$$

$$\left.\begin{array}{l} LA/pTA\ k_L L = n\pi,\ n\ \text{odd} \\ pTA/LA\ k_T L = n\pi,\ n\ \text{even} \end{array}\right\} \text{eqn (6.10)}$$

where n_L and n_T are integers. Otherwise, as mentioned before, solutions have to be found numerically.

Energy normalization gives $A = \left(\frac{\hbar}{K^2 2M\omega}\right)^{1/2} (a + a^\dagger)$, where for the antisymmetric wave:

$$K_A^2 = \frac{1}{2}\left[K_A^2(k_L) + K_A^2(k_T) + K_A^2(k_L, k_T)\right] \tag{6.17a}$$

$$K_A^2(k_L) = k_L^2 + k_x^2 - (k_L^2 - k_x^2)\frac{\sin k_L L}{k_L L} \tag{6.17b}$$

$$K_A^2(k_T) = \chi_A\left(k_T^2 + k_x^2 + (k_T^2 - k_x^2)\frac{\sin k_T L}{k_T L}\right) \tag{6.17c}$$

$$K_A^2(k_L, k_T) = \frac{4k_x}{L}\chi_A \cos(k_L L/2) \sin(k_T L/2) \tag{6.17d}$$

$$\chi_A = -\frac{r\,\sin(k_L L/2)}{t\,\sin(k_T L/2)} \tag{6.17e}$$

For the symmetric wave:

$$K_S^2 = \frac{1}{2}[K_S^2(k_L) + K_S^2(k_T) + K_S^2(k_L, k_T)] \tag{6.18a}$$

$$K_S^2(k_L) = \chi_S\left(k_L^2 + k_x^2 + (k_L^2 - k_x^2)\frac{\sin k_L L}{k_L L}\right) \tag{6.18b}$$

$$K_S^2(k_T) = \chi_S\left(k_T^2 + k_x^2 - (k_T^2 - k_x^2)\frac{\sin k_T L}{k_T L}\right) \tag{6.18c}$$

$$K_S^2(k_L, k_T) = -\frac{4k_x}{L}\chi_S \sin(k_L L/2) \cos(k_T L/2) \tag{6.18d}$$

$$\chi_S = -\frac{r\,\cos(k_L L/2)}{t\,\cos(k_T L/2)} \tag{6.18e}$$

6.2.2.2 Surface Modes

There is no surface mode for sTA, but equations (6.7) and (6.10) allow k_L and k_T to be purely imaginary: $k_L = i\alpha_L$, $k_T = i\alpha_T$. The mode patterns are antisymmetrical:

$$u_x = Ae^{ik_x x}\left(-k_x \sinh \alpha_L z + i\alpha_T \frac{r\cosh(\alpha_L L/2)}{t\cosh(\alpha_T L/2)} \sinh \alpha_T z\right)$$
$$u_z = iAe^{ik_x x}\left(i\alpha_L \cosh \alpha_L z + k_x \frac{r\cosh(\alpha_L L/2)}{t\cosh(\alpha_T L/2)} \cosh \alpha_T z\right)$$
(6.19)

and symmetrical:

$$u_x = Ae^{ik_x x}\left(k_x \cosh \alpha_L z + i\alpha_T \frac{r\sinh(\alpha_L L/2)}{t\sinh(\alpha_T L/2)} \cosh \alpha_T z\right)$$
$$u_z = iAe^{ik_x x}\left(-i\alpha_L \sinh \alpha_L z + k_x \frac{r\sinh(\alpha_L L/2)}{t\sinh(\alpha_T L/2)} \sinh \alpha_T z\right)$$
(6.20)

The secular equations, equations (6.7) and (6.10), become, respectively,

$$\tanh(\alpha_L L/2) = \frac{qr}{pt} \tanh(\alpha_T L/2)$$
$$\coth(\alpha_L L/2) = \frac{qr}{pt} \coth(\alpha_T L/2)$$
(6.21)

A simple solution can be found in the limit $L \to \infty$:

$$pt - qr = 0 \qquad (6.22)$$

This becomes a cubic equation for the slab velocity:

$$s^3 - 8s^2 + 8(3 - 2\gamma)s - 16(1 - \gamma) = 0 \qquad (6.23)$$

This is the equation found by Lord Rayleigh (1877) for acoustic surface waves (appropriately known as Rayleigh waves). An approximate value for the ratio of dilatation and shear coefficients that is generally applicable is $\gamma = 1/3$. If this is used, equation (6.23) reduces to

$$(s-4)(3s^2 - 12s + 8) = 0 \qquad (6.24)$$

and the solutions are

$$s = 4, \quad s = 2(1 + 1/\sqrt{3}), \quad s = 2(1 - 1/\sqrt{3}) \qquad (6.25)$$

The last solution ($s \sim 0.85$) is the only one consistent with k_{LT} imaginary. Thus, Rayleigh waves propagate more slowly than pTA waves.

For $k_x L \to 0$, the antisymmetric surface mode disappears. In the case of the symmetric mode, the pTA component converts to a guided wave with a real wave vector leaving the pLA as the only surface wave corresponding to a slab velocity:

$$s = 4(1 - \gamma), \quad \alpha_L^2 = (1 - 4\gamma + 4\gamma^2)k_x^2 \qquad (6.26)$$

Energy normalization gives the same form as equations (6.17) and (6.18) with suitable conversion of the wave vectors from real to imaginary.

6.3 Optical Modes

Applying the boundary condition $\mathbf{u} = 0$ at $z = \pm L/2$ leads to the mode pattern for the sTO mode:

$$u_y = A e^{ik_x x} \begin{cases} \cos k_z z, & k_z L = n\pi, \ n \text{ odd} \\ \sin k_z z, & k_z L = n\pi, \ n \text{ even} \end{cases} \qquad (6.27)$$

for which, as in the acoustic case,

$$K^2 = \frac{1}{2} \qquad (6.28)$$

The double hybrid pattern for the p modes is

$$\begin{aligned} u_x &= A e^{ik_x x} k_x (\sin k_L z - \lambda_a \sin k_T z) \\ u_z &= i A e^{ik_x x} \left(-k_L \cos k_L z - \frac{k_x^2}{k_T} \lambda_a \cos k_T z \right) \\ \lambda_a &= \frac{\sin(k_L L/2)}{\sin(k_T/2)} \end{aligned} \qquad (6.29)$$

These are antisymmetric modes that obey

$$\cot(k_L L/2) = -(k_x^2/k_L k_T) \cot(k_T L/2) \qquad (6.30)$$

The symmetric solution is

$$\begin{aligned} u_x &= A e^{ik_x x} k_x (\cos k_L z - \lambda_s \cos k_T z) \\ u_z &= i A e^{ik_x x} \left(k_L \sin k_L z - \frac{k_x^2}{k_T} \lambda_s \sin k_T z \right) \\ \lambda_s &= \frac{\cos(k_L L/2)}{\cos(k_T L/2)} \end{aligned} \qquad (6.31)$$

They obey

$$\tan(k_L L/2) = -(k_x^2/k_L k_T) \tan(k_T L/2) \qquad (6.32)$$

Note that the wave vectors of the LO and TO components are related by the frequency condition:

$$\omega^2 = \omega_L^2 - v_L^2(k_x^2 + k_L^2) = \omega_T^2 - v_T^2(k_x^2 + k_T^2) \qquad (6.33)$$

LO and TO hybrids are distinguished by frequency.

As for the acoustic case, when k_x is small, the dispersion is satisfied by

$$\left.\begin{array}{l} k_L L = n\pi, \ n \ even \\ k_T L = n\pi, \ n \ odd \end{array}\right\} \text{eqn (6.29)}$$
$$\left.\begin{array}{l} k_L L = n\pi, \ n \ odd \\ k_T L = n\pi, \ n \ even \end{array}\right\} \text{eqn (6.31)}$$
(6.34)

Energy normalization gives the K factors; the symmetric solution is

$$K_S^2 = \frac{1}{2}[K_S^2(k_L) + K_S^2(k_T) + K_S^2(k_L, k_T)] \tag{6.34a}$$

$$K_S^2(k_L) = k_L^2 + k_x^2 + (k_L^2 - k_x^2)\frac{\sin k_L L}{k_L L} \tag{6.34b}$$

$$K_S^2(k_T) = \left(\frac{k_x}{k_T}\right)^2 \lambda_S^2 \left(k_x^2 + k_T^2 + (k_x^2 - k_T^2)\frac{\sin k_T L}{k_T L}\right) \tag{6.34c}$$

$$K_S^2(k_L, k_T) = 8\frac{k_x^2}{k_T L}\lambda_S \sin(k_L L/2) \cos(k_T L/2) \tag{6.34d}$$

The antisymmetric solution is

$$K_A^2(k_L) = k_L^2 + k_x^2 - (k_L^2 - k_x^2)\frac{\sin k_L L}{k_L L} \tag{6.35a}$$

$$K_A^2(k_T) = \left(\frac{k_x}{k_T}\right)^2 \lambda_A^2 \left(k_x^2 + k_T^2 - (k_x^2 - k_T^2)\frac{\sin k_T L}{k_T L}\right) \tag{6.35b}$$

$$K_S^2(k_L, k_T) = 8\frac{k_x^2}{k_T L}\lambda_S \cos(k_L L/2) \sin(k_T L/2) \tag{6.35c}$$

There are no surface optical modes in non-polar material. Guided LO modes and evanescent electromagnetic modes in a polar slab were described by Fuchs and Kliewer (1965), in which the slab was defined as a dielectric continuum (DC). Mechanical and elastic boundary conditions were not considered. Rather than treat the polar slab with the full set of boundary conditions, we prefer to focus on the more technologically important quantum well.

7
Single Heterostructure

7.1 The Hybrid Model for Polar Optical Modes

The single heterostructure is one of the most technologically versatile devices, being the structure of field-effect transistors and Schottky-effect devices, to say nothing of its capability of exhibiting the quantum Hall effect at low temperatures. Here, we will focus on a heterostructure composed of III-V compounds such as AlAs/GaAs at room temperature and above, where optical waves are readily excited. We choose a particularly simple geometry in which the interface between two binary compounds is perpendicular to the z direction, and the structure is bounded in the z direction at $\pm L$ with the interface at $z = 0$ (Fig. 7.1). Anticipating the need to consider the electron–phonon interaction, we label the binary at $-L \geq z \leq 0$ as the barrier, and the binary at $0 \geq z \leq L$ as the well.

We will continue to use the distinction between sTA and pTA modes (Chapter 6, Fig. 6.1). The interface is in the xy plane perpendicular to z. sTA modes are polarized along the y direction, and pTA modes have polarization components along the x and z directions. Sometimes the sTA and pTA modes are referred to as TE and TM modes respectively, corresponding to electromagnetic waves, the former having the electric field polarized along the y direction and the latter having the magnetic field polarized along the y direction.

Focusing on the LO mode in the GaAs-like material, which we will refer to as the well, we can express the x and z components of the displacement as a sum of LO, IF, and pTO modes, all with the same frequency, as determined by the LO dispersion, assumed to be of parabolic form. The wave vectors of the components are

$$k_{LO}^2 = \frac{\omega_L^2 - \omega^2}{v_L^2} - k_x^2$$

$$k_{TO}^2 = \frac{\omega_T^2 - \omega^2}{v_T^2} - k_x^2 (= -\eta^2) \tag{7.1}$$

$$k_{IF}^2 = \frac{\omega^2}{c^2} - k_x^2 (\approx -k_x^2)$$

Hybrid Phonons in Nanostructures. First Edition. B.K. Ridley. © B.K. Ridley 2017.
Published in 2017 by Oxford University Press. DOI: 10.1093/acprof:oso/9780198788362.001.0001

70   Single Heterostructure

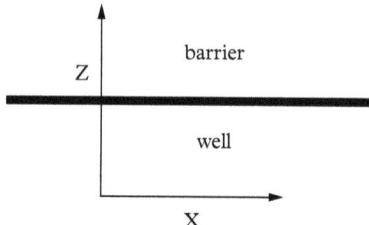

Figure 7.1 *The single heterostructure.*

Here, $v_{L,T}$ are velocities defining the lattice dispersion; c is a much larger velocity defining the electromagnetic dispersion.

That it is possible for the components to have the same frequency is the result of including dispersion for the optical modes (Figs 7.2 and 7.3). Writing for the evanescent transverse mode $\omega^2 = \omega_T^2 - v_T^2(k_x^2 - \eta^2)$ is pushing the long-wavelength parabolic approximation far beyond its validity, unless the frequency is close to the transverse frequency. In most cases of interest, the frequency will be that depicted in equation (7.1), that is, near the LO frequency, in which case a full-blooded calculation would be required to determine the magnitude of the imaginary wave vector that determines the form of the evanescent TO mode. Fortunately, it is usually sufficient that its magnitude be very large.

The resultant triple hybrid then has the form

$$u_x = e^{i(k_x x - \omega t)} \left(k_x A \cos k_z z + k_x B \sin k_z z + k_x C e^{-k_x z} + \eta D e^{-\eta z} \right)$$
$$u_z = i e^{i(k_x x - \omega t)} \left(k_z A \sin k_z z - k_z B \cos k_z z + k_x C e^{-k_x z} + k_x D e^{-\eta z} \right) \quad (7.2)$$

Here, A and B are the amplitudes of the incident and reflected LO waves; C and D are the amplitudes of the IF and TO evanescent waves. The choice of wave

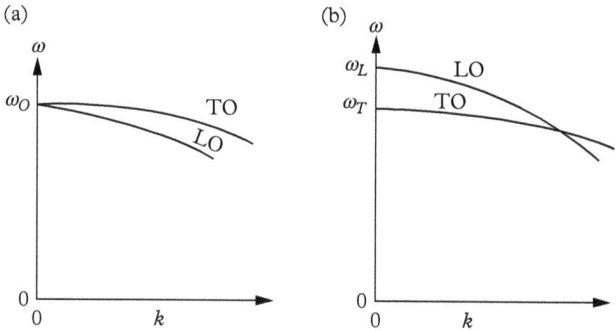

Figure 7.2 *General forms of LO and TO dispersion: (a) non-polar and (b) polar.*

The Hybrid Model for Polar Optical Modes

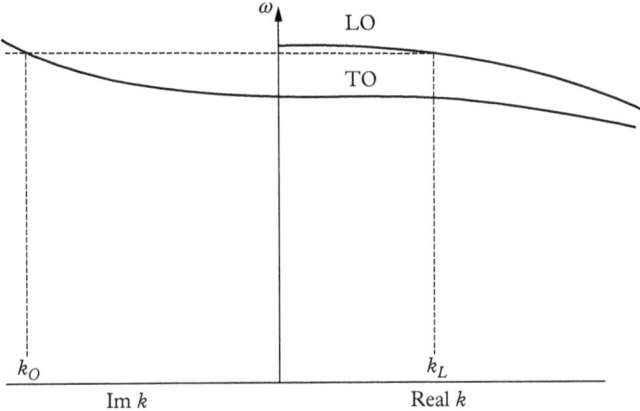

Figure 7.3 *LO and TO dispersion.*

vectors that multiply the amplitudes have been chosen so that the LO component has vanishing curl and the other components, being transverse, have vanishing divergences. Accompanying is an electrical wave whose field components are proportional to the displacements, as described in Section 3.2 (equations (3.8) and (3.18)):

$$E_x = -\alpha_0 e^{i(k_x x - \omega t)} \left[k_x (A \cos k_z z + B \sin k_z z) + s k_x C e^{-k_x z} \right]$$
$$E_z = -i\alpha_0 e^{i(k_x x - \omega t)} \left[k_z (A \sin k_z z - B \cos k_z z) + s k_x C e^{-k_x z} \right] \quad (7.3)$$

$$\alpha_0 = \frac{e^*}{V_0 \varepsilon_\infty}, \quad s = \frac{\omega^2 - \omega_T^2}{\omega_L^2 - \omega_T^2} \quad (7.4)$$

Here, s is known as the field factor. In what follows we suppress the frequency factor $e^{-i\omega t}$. We assume the mechanical boundary condition $\mathbf{u} = 0$ at $z = 0$.

In addition, there is an electric field in the barrier:

$$E_x^B = k_x F e^{ik_x x + k_x z}$$
$$E_z^B = -i k_x F e^{ik_x x + k_x z} \quad (7.5)$$

($z < 0$), connected to the electric field in the well by the usual electric boundary conditions (E_x and D_z continuous):

$$-\alpha_0 [A + sC] = F$$
$$\varepsilon_W(\omega) \alpha_0 s C = \varepsilon_B(\omega) F \quad (7.6)$$

Equation (7.6) takes into account that the electric displacement of the LO mode is zero. $\varepsilon_{W,B}(\omega)$ are the transverse permittivities of the well and barrier (equation (3.18)). Solving equations (7.2) and (7.6) for the amplitudes, we obtain

$$u_x = k_x A e^{ik_x x} \left[\cos k_z z - \Gamma \sin k_z z - pe^{-k_x z} - [(1-p)/\eta]e^{-\eta z} \right]$$
$$u_z = ik_z A e^{ik_x x} \left[\sin k_z z + \Gamma \cos k_z z - (k_x/k_z)pe^{-k_x z} - (k_x^2/k_z^2)[(1-p)/\eta]e^{-\eta z} \right] \quad (7.7)$$

$$\Gamma = (k_x/k_z)p[1 + (k_x/\eta p)(1-p)]$$

$$p = [s(1+r)]^{-1}, \quad r = \frac{\varepsilon_W(\omega)}{\varepsilon_B(\omega)} \quad (7.8)$$

The electric field in the barrier is

$$E_x^B = -\alpha_0 srp k_x A e^{ik_x x + k_x z}$$
$$E_z^B = i\alpha_0 srp k_x A e^{ik_x x + k_x z} \quad (7.9)$$

The analysis can be repeated for the optical modes in the barrier. Just as an electric field associated with the optical modes in the well appears in the barrier, a corresponding field associated with the optical modes in the barrier appears in the well.

It is worth noting that in the absence of lattice dispersion $s = 1$, $p = 1$, and $r = 0$, and the barrier fields disappear.

7.2 Remote Phonons

Because of lattice dispersion, electric fields at the well frequency appear in the barrier and, equally, electric fields at the barrier frequency appear in the well. These fields are responsible for so-called remote-phonon scattering (Wang and Mahan 1972). Besides their effect on scattering, these remote-phonon fields are associated with corresponding lattice displacements according to equations (3.8) and (3.18), which has consequences for the connection rules at the interface. Thus, for example, the fields depicted in equation (7.9) will be associated with displacement components u_x and u_z in the barrier, which are non-zero. This violates the condition **u** = 0 at the interface. The condition is restored provided that the IF mode in the barrier is a component of a hybrid of LO, TO and IF modes at the well frequency. A corollary of our assumption that **u** = 0 at the interface is that there is a large disparity between the frequencies of well and barrier; consequently the LO and TO components will be heavily evanescent, with exponents derived from the optical-acoustic complex band structure. The barrier hybrid at the well frequency then has the form ($z < 0$)

$$u_x = e^{i(k_x x - \omega t)} \left(k_x P e^{\eta_L z} + \eta_T Q e^{\eta_T z} + k_x R e^{k_x z} \right)$$
$$u_z = -i e^{i(k_x x - \omega t)} \left(\eta_L P e^{\eta_L z} + k_x Q e^{\eta_T z} + k_x R e^{k_x z} \right) \quad (7.10)$$

Solving for $u_x = 0$ and $u_z = 0$ leads to

$$u_x = k_x R e^{i(k_x x - \omega t)} \left(-p_L \frac{1 - p_T}{1 - p_L p_T} e^{\eta_L z} - \frac{1 - p_L}{1 - p_L p_T} e^{\eta_T z} + e^{k_x z} \right)$$
$$u_x = i k_x R e^{i(k_x x - \omega t)} \left(-\frac{1 - p_T}{1 - p_L p_T} e^{\eta_L z} + p_T \frac{1 - p_L}{1 - p_L p_T} e^{\eta_T z} - e^{k_x z} \right) \quad (7.11)$$

In these equations:

$$p_L = \frac{k_x}{\eta_L}, \; p_T = \frac{k_x}{\eta_T} \quad (7.12)$$

As regards the effect on the electrical connection rules, the TO component has no effect, while the large magnitude of η_L make the contribution to the tangential field negligible. The electric fields depicted in equation (7.9) are unchanged.

7.3 Energy Normalization

In the process of energy normalization, the amplitude A is related to the operator $\hat{q}$ acting on the number states of the phonon:

$$\hat{q} = \left(\frac{\hbar}{2 \overline{M} \omega} \right)^{1/2} (a_{k_x} + a^\dagger_{-k_x}) e^{i k_x x} \quad (7.13)$$

where $\overline{M}$ is the reduced mass. The mechanical energy of the hybrid is

$$U = \overline{M} \omega^2 \int_0^L (|u_x|^2 + |u_z|^2) dz / L = \overline{M} \omega^2 |\hat{q}|^2 \quad (7.14)$$

$$A_{\mathbf{k}} = \left(\frac{\hbar}{2 \mu \omega K^2} \right)^{1/2} (a_{k_x} + a^\dagger_{-k_x}) e^{i k_x x} \quad (7.15)$$

$$K^2 = \frac{1}{2} \left[\begin{array}{c} (k_x^2 + k_z^2)(1 + \Gamma^2) + \frac{2 p^2 k_x^2}{k_x L} + \frac{(1-p)^2 k_x^2}{\eta^3 L}(\eta^2 + k_x^2) \\ -\frac{4 p k_x}{L} - \frac{4(1-p) k_x^2}{(\eta^2 + k_z^2) L}(\eta + k_z) + \frac{p(1-p) k_x^2}{\eta L} \end{array} \right] \quad (7.16)$$

We have used the periodic boundary condition $k_z L = 2\pi n$ and have assumed that L is large enough for $\exp(-k_x L) \to 0$, $\exp(-\eta L) \to 0$. In most cases of interest

74 Single Heterostructure

L is large enough for the terms inversely proportional to L to be ignored, in which case

$$K^2 = \frac{1}{2}(k_x^2 + k_z^2)(1 + \Gamma^2) \tag{7.17}$$

We note that the fields induced in the barrier produce optical displacements in the barrier. Thus, part of the total mechanical energy at the frequency in the well resides in the barrier and has to be taken into account in the energy normalization. With L large, however, the contribution is negligible.

7.4 Reduced Boundary Condition

As mentioned previously, the disparity between the TO and LO frequencies ensures that η is very large in most cases of interest, which enables us to ignore the role of the TO mode in satisfying the mechanical boundary condition for u_z. In this approximation, equations (7.7) and (7.8) become

$$\begin{aligned} u_x &= k_x A e^{ik_x x}\left[\cos k_z z - \Gamma \sin k_z z - p e^{-k_x z}\right](1 - \delta_{z,0}) \\ u_z &= i k_z A e^{ik_x x}\left[\sin k_z z + \Gamma \cos k_z z - (k_x/k_z) p e^{-k_x z}\right] \end{aligned} \tag{7.18}$$

$$\Gamma = (k_x/k_z) p$$

$$p = [s(1 + r)]^{-1}, \quad r = \frac{\varepsilon_W(\omega)}{\varepsilon_B(\omega)} \tag{7.19}$$

Energy normalization reduces K^2 to

$$K^2 = \frac{1}{2}\left[(k_x^2 + k_z^2)(1 + \Gamma^2) + \frac{2p^2 k_x^2}{k_x L} - \frac{4 p k_x}{L}\right] \tag{7.20}$$

and thence to equation (7.17) in the limit L tending to ∞.

With the electron–phonon interaction in mind we write down the electric potential corresponding to the displacements of equation (7.18) that the electron in the well experiences:

$$\phi = -i\alpha_0 A e^{ik_x x}\left[\cos k_z z - \Gamma \sin k_z z - s p e^{-k_x z}\right] \tag{7.21}$$

($\mathbf{E} = -\nabla\phi$). There will also be an interaction of electrons in the well with the induced interface fields in the well at the frequency near the LO frequency in the barrier.

The approximation can be pushed further if only extremely long wavelengths are involved, as in some cases of Raman spectroscopy. In this case

$$\omega \approx \omega_L, \; s \approx 1 \text{ and } \varepsilon_W(\omega) \approx 0 \tag{7.22}$$

and hence $r = 0$. This is equivalent to neglecting lattice dispersion entirely. The displacements and barrier fields reduce to

$$u_x = k_x A e^{ik_x x} \left[\cos k_z z - (k_x/k_z) \sin k_z z - e^{-k_x z} \right]$$
$$u_z = ik_z A e^{ik_x x} \left[\sin k_z z + (k_x/k_z) \cos k_z z - (k_x/k_z) e^{-k_x z} \right] \quad (7.23)$$

It is worth noting that in the absence of lattice dispersion the mechanical boundary conditions can be satisfied without the assistance of the TO mode.

7.5 Acoustic Hybrids

Acoustic modes that encounter the interface, unlike optical modes, can be partly transmitted into the barrier, there being common acoustic frequencies in both well and barrier. The usual boundary conditions—continuity of stress and displacement—must be satisfied. We continue to regard the material forming the heterojunction is being elastically isotropic, so that we can regard the TA and LA modes as decoupled in the bulk. (The case of elastic anisotropy has been described by Auld (1990). The TA modes divide into two sorts depending on their polarization relative to the interface. Those with polarization in the plane of the interface we denote as sTA modes; those with a component of polarization perpendicular to the plane of the interface we denote as pTA modes. By their very nature LA modes are always pLA modes. We need to consider three cases of an acoustic mode incident on the interface: (1) an sTA, (2) a pTA, and (3) a pLA mode.

We will assume that the piezoelectricity is comparatively weak ($K^2 \ll 1$) and can be ignored in determining the mechanical boundary conditions.

7.5.1 sTA

We take the polarization to lie in the y direction. Thus, the displacement can be taken to be

$$u_y = e^{ik_x x} \left(A e^{ik_1 z} + B e^{-ik_1 z} \right) \quad (7.24)$$

where B is the amplitude of the reflected wave. It has the zero divergence of a transverse mode. We consider that there exists a transmitted wave in the barrier of form

$$u_y = e^{ik_x x} C e^{ik_2 z} \quad (7.25)$$

The wave vector in the well is k_1 and the wave vector in the barrier is k_2. The time dependence in equations (7.24) and (7.25) has been suppressed. The transmitted wave must have the same frequency as that of the incident wave. If v_1 is the velocity of transverse waves in the well and v_2 is the velocity of transverse waves in the barrier, then

$$\omega^2 = v_1^2 \left(k_x^2 + k_1^2\right) = v_2^2 \left(k_x^2 + k_2^2\right)$$
$$\therefore k_2^2 = k_x^2(v-1) + vk_1^2, v = v_1^2/v_2^2 \tag{7.26}$$

The stress that must be continuous at $z = 0$ is

$$T_4 = c_{44} \frac{\partial u_y}{\partial z}$$
$$c_{44}^1 i k_1 (A - B) = c_{44}^2 i k_2 C \tag{7.27}$$

Continuity of displacement implies that

$$A + B = C \tag{7.28}$$

For simplicity, we will assume the (so-called) mass approximation and take $c_{44}^1 = c_{44}^2$. (This approximation assumes that the difference in wave velocity is primarily due to difference in density.) Thus

$$B = A \frac{k_1 - k_2}{k_1 + k_2}, \quad C = A \frac{2k_1}{k_1 + k_2} \tag{7.29}$$

We notice that when $v_2^2 > v_1^2$, $k_2^2 = 0$ when $k_x^2 = [v/(1-v)]k_1^2$. This condition corresponds to total internal reflection at an angle of incidence given by Snell's law:

$$\frac{\sin \theta_r}{\sin \theta_i} = \frac{v_2}{v_1}, \quad \therefore \sin \theta_{tir} = (v_1/v_2) \tag{7.30}$$

$$\sin \theta_i = \left(\frac{k_x^2}{k_x^2 + k_1^2}\right)^{1/2}, \quad \sin \theta_r = \left(\frac{k_x^2}{k_x^2 + k_2^2}\right)^{1/2} \tag{7.31}$$

Within the mass approximation, the condition applies when the barrier material is less dense than the well material, as is the case, for example, for the system AlAs/GaAs.

7.5.2 pTA

Things are not so straightforward for the p-polarized TA mode, which has to satisfy the continuity of two stress components as well as the continuity of the displacement components. In order to do this it has to combine with a pLA mode in the well:

$$u_x = e^{ik_x x}(k_T A e^{ik_T z} + k_T B e^{-ik_T z} + k_x C e^{-ik_L z})$$
$$u_z = e^{ik_x x}(-k_x A e^{ik_T z} + k_x B e^{-ik_T z} - k_L C e^{-ik_L z}) \qquad (7.32)$$

and the refracted beam has to combine with a pLA mode in the barrier:

$$u_x = e^{ik_x x}(q_T D e^{iq_T z} + k_x E e^{iq_L z})$$
$$u_z = e^{ik_x x}(-k_x D e^{ik_2 z} + q_L E e^{iq_L z}) \qquad (7.33)$$

The inclusion of wave vectors along with the amplitudes ensures that the TA component has no divergence and the LA mode has no curl. The frequency of each component has to be the same.

The relevant stress components that must be continuous are

$$T_3 = c_{11} \nabla \cdot \mathbf{u} - 2 c_{44} \frac{\partial u_x}{\partial x}$$
$$T_5 = c_{44} \left(\frac{\partial u_z}{\partial x} + \frac{\partial u_x}{\partial z} \right) \qquad (7.34)$$

In the mass approximation, c_{11} and c_{44} are the same in well and barrier, assuming that this the case, continuity of stress and displacement leads to the equations

$$-2\gamma k_x k_T B + [k_x^2(1-2\gamma) + k_L^2]C + 2\gamma k_x q_T D - [k_x^2(1-2\gamma) + q_L^2]E = 2\gamma k_x k_T A$$
$$(k_x^2 - k_T^2)B - 2k_x k_L C + (k_x^2 - q_T^2)D - 2k_x q_L E = (k_x^2 - k_T^2)A \qquad (7.35)$$
$$k_T B + k_x C - q_T D - k_x E = -k_T A$$
$$k_x B - k_L C + k_x D - q_L E = k_x A$$

Here, $\gamma = c_{44}/c_{11}$. These equations have solutions

$$B = A \frac{\Delta_B}{\Delta}, \quad C = A \frac{\Delta_C}{\Delta}, \quad D = A \frac{\Delta_D}{\Delta}, \quad E = A \frac{\Delta_E}{\Delta} \qquad (7.36)$$

Here, Δ is the determinant of the coefficients on the left-hand side of equation (7.35), Δ_B is the determinant with the coefficients of A replacing those of B, similarly for C, D, and E.

The amplitudes can be obtained numerically, but it is useful to consider the case of nearly normal incidence in which approximate results for the amplitudes can be obtained. At normal incidence $k_x = 0$ and

$$B = A \frac{k_T - q_T}{k_T + q_T}, \quad D = A \frac{2 k_T^2}{q_T(k_T + q_T)}, \quad C = E = 0 \qquad (7.37)$$

No hybridization is necessary. Approximate values for the LA components can be found for $k_x \sim 0$ using the values for B and D given in equation (7.37):

$$C = -\frac{2k_x q_T q_L}{k_L(k_L + q_L)(k_T + q_T)}A, \quad E = -\frac{2k_x q_T k_L}{q_L(k_L + q_L)(k_T + q_T)}A \tag{7.38}$$

These amplitudes increase with angle of incidence, linearly at first. Ultimately, their associated wave vectors become purely imaginary and the LA components become evanescent waves at certain critical angles. Taking the common case to be such that the wave velocities obey $v_{L2} > v_{L1} > v_{T2} > v_{T1}$, we can identify three critical angles or, what is the equivalent thing, critical values of k_x. The first is associated with the total internal reflection of the LA component and heralds the conversion of the LA component in the barrier into an evanescent wave:

$$k_x^2 = \frac{v_{T1}^2 k_T^2}{v_{L2}^2 - v_{T1}^2}, \quad q_L = 0 \tag{7.39}$$

The second corresponds to

$$k_x^2 = \frac{v_{T1}^2 k_T^2}{v_{L1}^2 - v_{T1}^2}, \quad k_L = 0 \tag{7.40}$$

The third corresponds to the total internal reflection of the TA component:

$$k_x^2 = \frac{v_{T1}^2 k_T^2}{v_{T2}^2 - v_{T1}^2}, \quad q_T = 0 \tag{7.41}$$

It is of interest to notice that the LA component excited in the well at the interface by the TA wave introduces an interaction with electrons via the deformation potential that would otherwise be absent (ignoring the weak piezoelectric interaction) in zinc blende materials.

7.5.3 pLA

There is a closely similar pattern for the incident pLA mode: reflected and transmitted LA modes accompanied by TA modes in the well and barrier. The components of the hybrid in the well have the structure:

$$\begin{aligned} u_x &= e^{ik_x x}(k_x A e^{ik_L z} + k_x B e^{-ik_L z} + k_T C e^{-ik_T z}) \\ u_z &= e^{ik_x x}(k_L A e^{ik_L z} - k_L B e^{-ik_L z} + k_x C e^{-ik_T z}) \end{aligned} \tag{7.42}$$

Here, A and B are the amplitudes of the incident and reflected LA waves and C is the amplitude of the TA induced component. In the barrier, we have

$$\begin{aligned} u_x &= e^{ik_x x}(k_x D e^{iq_L z} + q_T E e^{-iq_T z}) \\ u_z &= e^{ik_x x}(q_L D e^{iq_L z} + k_x E e^{-iq_T z}) \end{aligned} \tag{7.43}$$

Here, D and E are the amplitudes of the transmitted LA wave and the induced TA mode. Entailing the continuity of the stresses T_3 and T_5 and of the displacements defines the amplitudes, as in the case of the pTA mode.

Once more adopting the mass approximation, we obtain the amplitudes for the case of normal incidence:

$$B = -A\frac{k_L - q_L}{k_L + q_L}, \quad D = A\frac{2k_L}{(k_L + q_L)}, \quad C = E = 0 \tag{7.44}$$

For small angles of incidence, B and D are unchanged, and

$$C = -\frac{2k_x q_T (k_L - q_L)}{k_T (k_L + q_L)(k_T - q_T)} A$$
$$E = -\frac{2k_x k_T (k_L - q_L)}{q_T (k_L + q_L)(k_T - q_T)} A \tag{7.45}$$

There is now only one critical angle: that associated with total internal reflection. With the same hierarchy of velocities this corresponds to

$$k_x^2 = \frac{v_{L1}^2 k_L^2}{v_{L2}^2 - v_{L1}^2}, \quad q_L = 0 \tag{7.46}$$

The wave vectors of the TA modes are

$$q_T^2 = \frac{v_{L1}^2}{v_{T2}^2}(k_L^2 + k_x^2) - k_x^2$$
$$k_T^2 = \frac{v_{L1}^2}{v_{T1}^2}(k_L^2 + k_x^2) - k_x^2 \tag{7.47}$$

Neither mode becomes evanescent with increasing angle.

In all cases the amplitude of the incident wave can be expressed in terms of annihilation and creation operators via the usual process of energy normalization. In zinc blende material, the interaction with electrons proceeds through the deformation potential associated with the LA components and through the piezoelectric interaction with TA modes.

In the above, we have explicitly considered well modes incident on the interface, but there are also barrier modes that introduce LA and TA modes with barrier frequencies into the well. These barrier modes propagate in the well in directions that are quite different from those of the well mode of the same frequency. Electrons in the well interact with two distinct sorts of acoustic waves at a given frequency: the one deriving from the barrier, the other from the well.

7.5.4 General Remarks

In the case of the single heterostructure, what confinement exists for a travelling waves with real vectors is limited to a half-space. However, it is possible for interface modes to exist, known as Stonely waves (Stonely 1924).

7.6 Interface Acoustic Modes

These waves are pLA/TA hybrids with displacement components:

$$\begin{aligned} u_x &= e^{ik_x x}\left(\beta_T A e^{\beta_T z} + k_x B e^{\beta_L z}\right) \\ u_z &= e^{ik_x x}\left(-ik_x A e^{\beta_T z} + -i\beta_L B e^{\beta_L z}\right) \quad z \le 0 \end{aligned} \quad (7.48)$$

In the barrier, we have

$$\begin{aligned} u_x &= e^{ik_x x}\left(\eta_T C e^{-\eta_T z} + k_x D e^{-\eta_L z}\right) \\ u_z &= e^{ik_x x}\left(ik_x C e^{-\eta_T z} + i\eta_L D e^{-\eta_L z}\right) \quad z \ge 0 \end{aligned} \quad (7.49)$$

Satisfying the boundary conditions gives four equations for the amplitudes:

$$\begin{aligned} &\beta_T A + k_x B = \eta_T C + \eta_L D \\ &-k_x A - \beta_L B = k_x C + \eta_L D \\ &c^1_{11}\left[-2\gamma_1 k_x \beta_T A + \left[k_x^2(1-2\gamma_1) - \beta_L^2\right]B\right] = c^2_{11}\left[-2\gamma_2 k_x \eta_T C + \left[k_x^2(1-2\gamma_2) - \eta_L^2\right]D\right] \\ &c^1_{11}\gamma_1\left[(\beta_T^2 + k_x^2)A + 2k_x\beta_L B\right] = c^2_{11}\gamma_2\left[(\eta_T^2 + k_x^2)C + 2k_x\eta_L D\right] \end{aligned} \quad (7.50)$$

The superscript (subscript) 1, 2 refers to the well and barrier respectively. C_{11} is the elastic constant, and

$$\gamma = c_{44}/c_{11} \quad (7.51)$$

Note that we do not invoke the mass approximation here. It is convenient at this point to introduce the wave velocity **v** defined by

$$\omega = vk_x \quad (7.52)$$

and the ratio

$$s = v^2/v_T^2 \quad (7.53)$$

where v_T is the velocity of the TA mode.

A solution exists provided the following equation is satisfied:

$$(d_2 g_2 + f_2 e_2)(1 - a_1 b_1) + (e_1 g_2 + g_1 e_2)(a_1 + a_2) - (e_1 f_2 - g_1 d_2)(1 + a_1 b_2) - \\ (-d_1 g_2 + f_1 e_2)(1 + a_2 b_1) - (d_1 f_2 + f_1 d_2)(b_1 + b_2) + (d_1 g_1 + f_1 e_1)(1 - a_2 b_2) = 0 \quad (7.54)$$

where

$$\begin{aligned}
a_{1,2} &= (1 - s_{1,2})^{1.2} \\
b_{1,2} &= (1 - \gamma_{1,2} s_{1,2})^{1/2} \\
d_{1,2} &= 2 c_{11}^{1,2} \gamma_{1,2} a_{1,2} \\
e_{1,2} &= c_{11}^{1,2} \gamma_{1,2} (s_{1,2} - 2) \\
f_{1,2} &= c_{11}^{1,2} \gamma_{1,2} (2 - s_{1,2}) \\
g_{1,2} &= 2 c_{11}^{1,2} \gamma_{1,2} b_{1,2}
\end{aligned} \quad (7.55)$$

Note that $c_{11}^{1,2} \gamma_{1,2} = v_{T_{1,2}}^2 \rho_{1,2}$, where ρ is the mass density. In a particular case the solution for the velocity of the interface mode, v, has to be found numerically.

Stonely showed that a solution always exists. His demonstration is neat and worth repeating. Let $\rho_2 = 0$; all terms with subscript 2 vanish and equation (7.52) becomes

$$(2 - s)^2 - 4(1 - s)^{1/2}(1 - \gamma_1 s)^{1/2} = 0 \quad (7.56)$$

and therefore

$$s_1^3 - 8 s_1^3 + 8(3 - 2\gamma_1) s_1 - 16(1 - \gamma_1) = 0 \quad (7.57)$$

But this is exactly the equation for Rayleigh waves (equation (6.23)). Given a large difference between the elastic coefficients or density of the two adjacent materials there will always be an interface wave.

Now look at an opposite situation in which the densities are different, but where the TA velocities are the same and the LA velocities are the same: $v_{T1} = v_{T2} = v_T$, $v_{L1} = v_{L2} = v_L$. In this case, $s_1 = s_2 = s$, and

$$\begin{aligned}
a_1 &= a_2 = a = (1 - s)^{1/2} \\
b_1 &= b_2 = b = (1 - \gamma s)^{1/2} \\
d_{1,2} &= 2 v_T^2 \rho_{1,2} a \\
e_{1,2} &= v_T^2 \rho_{1,2} (s - 2) \\
f_{1,2} &= v_T^2 \rho_{1,2} (2 - s) \\
g_{1,2} &= 2 v_T^2 \rho_{1,2} b
\end{aligned} \quad (7.58)$$

Equation (7.54) becomes

$$s^2[(\rho_1 - \rho_2)^2 - ab(\rho_1 + \rho_2)^2] - 4(\rho_1 - \rho_2)^2[(ab)^2 + 1 - s) + ab(s - 2)] = 0 \quad (7.59)$$

Label the left-hand side of equation (7.59) $F(s)$, so that equation (7.59) becomes $F(s) = 0$.

When $s = 1$, $F(1) = (\rho_1 - \rho_2)^2$, which is positive. When $s = 0$, $F(0) = 0$. Now let s be small but finite. It is useful to put $ab = ab - 1 + 1$. To lowest order,

$$ab - 1 = [1 - (1 + \gamma)s]^{1/2} - 1 \approx -\frac{1}{2}(1 + \gamma)s \quad (7.60)$$

and

$$F(s) = s^2[(\rho_1 - \rho_2)^2 \gamma^2 - (\rho_1 + \rho_2)^2] \quad (7.61)$$

It follows that when s is small, $F(s)$ is negative. But when $s = 1$, $F(s)$ is positive. Thus, somewhere in between, $F(s) = 0$ for finite s, and, therefore, an interface mode exists.

Single heterostructures support a rich variety of acoustic waves, ranging from unconfined modes, modes limited to a half space, and interface modes. An account of their interaction with electrons will need to take into account the special characteristics of each type.

8
Quantum Well

8.1 Triple Hybrid

We adopt the usual simplifying assumptions, namely:

(1) The long-wavelength approximation—decouples acoustic and optical modes.
(2) The crystals are isotropic—decouples LO and TO modes.
(3) The disparity of the optical mode frequencies across the interface is large, so that the appropriate boundary condition that achieves mechanical equilibrium is $\mathbf{u} = 0$.
(4) The properties of electrical interface modes are those determined near the LO frequency.
(5) Evanescent TO modes exist with frequency equal to the LO frequency (Figs 7.2 and 7.3).

We consider a slab of polar material bounded in the z direction at $z = \pm L/2$ and unbounded otherwise (Fig. 8.1). We take the boundaries to be of such material that entirely inhibits the transmission of polar modes (and of electrons, which are confined into quantized states in the z direction, hence the nomenclature, though our attention will be focused here entirely on optical modes). For simplicity, we assume that the quantum well has identical barriers, unbounded in the z direction, that are also elastically isotropic.

Once again, we need only to consider p-polarized LO and TO waves: the s-polarized TO wave can satisfy the usual boundary condition of $\mathbf{u} = 0$ without hybridization. Exploiting the symmetry of the quantum well, we can write the components of the lattice displacement, suppressing the time dependence, as follows:

$$u_x = e^{ik_x x}(k_x A \cos k_z z + k_x B \cosh k_x z + \eta C \cosh \eta z)$$
$$u_z = e^{ik_x x}(k_z A \sin k_z z - k_x B \sinh k_x z - k_x C \sinh \eta z) \quad (8.1)$$

Hybrid Phonons in Nanostructures. First Edition. B.K. Ridley. © B.K. Ridley 2017.
Published in 2017 by Oxford University Press. DOI: 10.1093/acprof:oso/9780198788362.001.0001

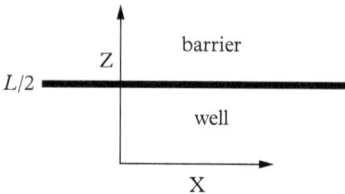

Figure 8.1 *The quantum well.*

The electric fields associated with these displacements are

$$E_x = -\alpha e^{ik_x x}(k_x A \cos k_z z + k_x s B \cosh k_x z)$$
$$E_z = -\alpha e^{ik_x x}(k_z A \sin k_z z - k_x s B \sinh k_x z) \quad (8.2a)$$

Here

$$\alpha = e^*/V_0 \varepsilon_\infty, \quad s = \frac{\omega^2 - \omega_T^2}{\omega_L^2 - \omega_T^2} \quad (8.2b)$$

with s being the field factor of the interface (IF) mode; $\omega_{L,T}$ the zone-centre LO and TO frequencies.

The induced fields in the barriers, assumed to be infinitely thick, are

$$\left.\begin{array}{l} E_x^B = k_x F e^{ik_x x} e^{k_x(z+L/2)} \\ E_z^B = -ik_x F e^{ik_x x} e^{k_x(z+L/2)} \end{array}\right\} z \leq -L/2 \quad \left.\begin{array}{l} E_x^B = k_x F e^{ik_x x} e^{-k_x(z+L/2)} \\ E_z^B = -ik_x F e^{ik_x x} e^{-k_x(z+L/2)} \end{array}\right\} z \geq L/2$$
$$(8.3)$$

There will also be fields induced from the barrier into the well. We will consider these fields later.

The mechanical and electrical boundary conditions can be satisfied at $z = \pm L/2$ by the following solution:

$$u_x = k_x A e^{ik_x x}[\cos k_z z - \Gamma_{s1} \cosh k_x z - \Gamma_{s2} \cosh \eta z]$$
$$u_z = ik_z A e^{ik_x x}\left[\sin k_z z + (k_x/k_z)\Gamma_{s1} \sinh k_x z + (k_x^2/k_z \eta)\Gamma_{s2} \sinh \eta z\right] \quad (8.4)$$

$$\Gamma_{s1} = \frac{p_s \cos(k_z L/2)}{\sinh(k_x L/2)}, \quad \Gamma_{s2} = \frac{\cos(k_z L/2)}{\cosh(\eta L/2)}[1 - p_s \coth(k_x L/2)]$$
$$p_s = \frac{1}{s[\coth(k_x L/2) + r]} \quad (8.5)$$

Here, the ratio of the permittivities is $r = \varepsilon_W(\omega)/\varepsilon_B(\omega)$ (subscripts W and B standing for well and barrier). The condition on k_z is given by

$$\tan(k_z L/2) = -(k_x/k_L)p_s - (k_x^2/k_z \eta)[1 - p_s \coth(k_x L/2)]\tanh(\eta L/2) \quad (8.6)$$

The fields induced in the barrier are

$$E_x^B = \pm iE_z^B = -k_x A r\alpha s p_s \cos(k_z L/2) e^{ik_x x \pm k_x(z \pm L/2)} \quad (8.7)$$

These results refer to the symmetric solution, defined, as before, by the pattern of the x component. The antisymmetrical solution is

$$\begin{aligned} u_x &= k_x A e^{ik_x x} [\sin k_z z - \Gamma_{a1} \sinh k_x z - \Gamma_{a2} \sinh \eta z] \\ u_z &= ik_z A e^{ik_x x} \left[\cos k_z z - (k_x/k_z)\Gamma_{a1} \cosh k_x z - (k_x^2/k_z \eta)\Gamma_{a2} \cosh \eta z \right] \end{aligned} \quad (8.8)$$

$$\Gamma_{a1} = \frac{p_a \sin(k_z L/2)}{\cosh(k_x L/2)}, \quad \Gamma_{a2} = \frac{\sin(k_z L/2)}{\sinh(\eta L/2)}[1 - p_a \tanh(k_x L/2)]$$
$$p_a = \frac{1}{s[\tanh(k_x L/2) + r]} \quad (8.9)$$

The condition on k_z is

$$\cot(k_z L/2) = (k_x/k_L) p_a + (k_x^2/k_z \eta)[1 - p_a \tanh(k_x L/2)] \coth(\eta L/2) \quad (8.10)$$
$$E_x^B = \pm i E_z^B = \mp k_x A r\alpha s p_a \sin(k_z L/2) e^{ik_x x \pm k_x(z \pm L/2)} \quad (8.11)$$

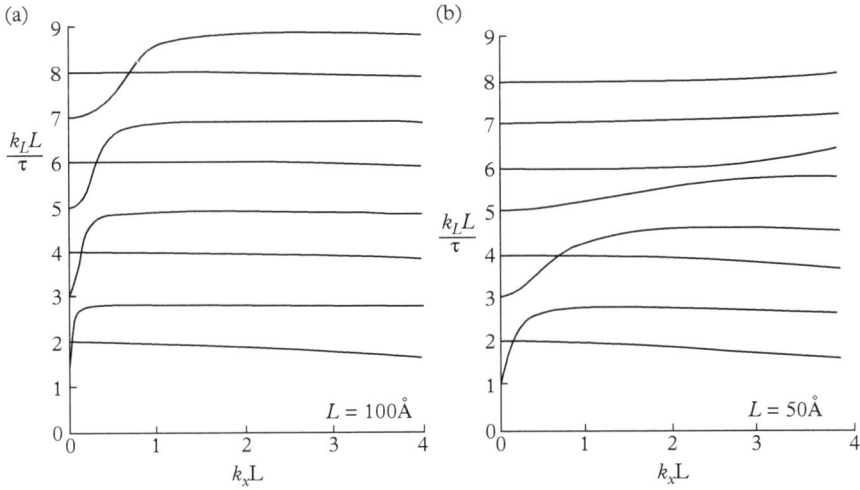

Figure 8.2 *Dispersion in a GaAs quantum well, well-width (a) 100 Å, (b) 50 Å. Reprinted with permission from Cambridge University Press.*

86 Quantum Well

An example of the resultant dispersion is shown in Figure 8.2. Antisymmetric modes exhibit changes with the in-plane wave vector from LO-like to IF-like and back again; the symmetric modes remain LO-like until $\coth(k_x L/2) + r \approx 0$. Mode patterns for $k_x \sim 0$ are shown in Figure 8.3, and for $k_x \sim 3$ in Figure 8.4. Patterns at LO to IF conversion are shown in Figure 8.5 for antisymmetric modes; the pattern for a symmetric mode is shown in Figure 8.6.

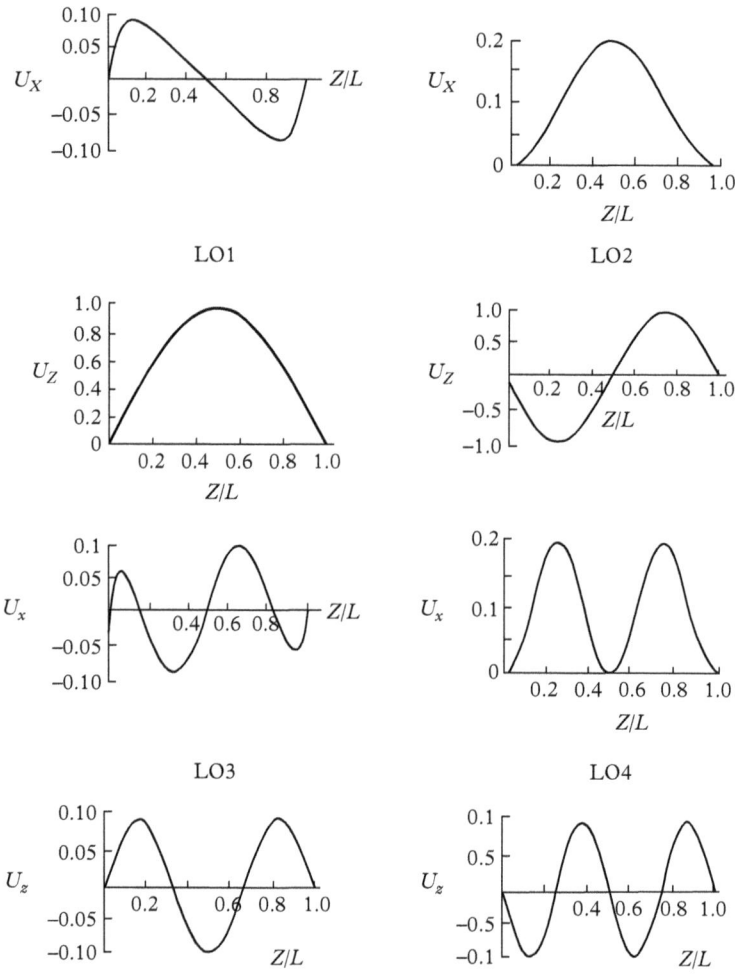

Figure 8.3 Mode patterns for $k_x \sim 0$ ($L = 38$ Å). Reprinted with permission from Cambridge University Press.

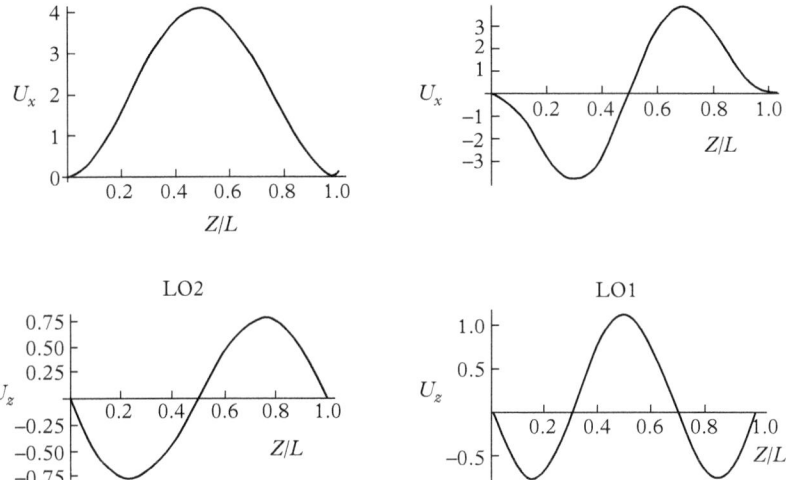

Figure 8.4 Mode patterns for $k_x L = 3$. Reprinted with permission from Cambridge University Press.

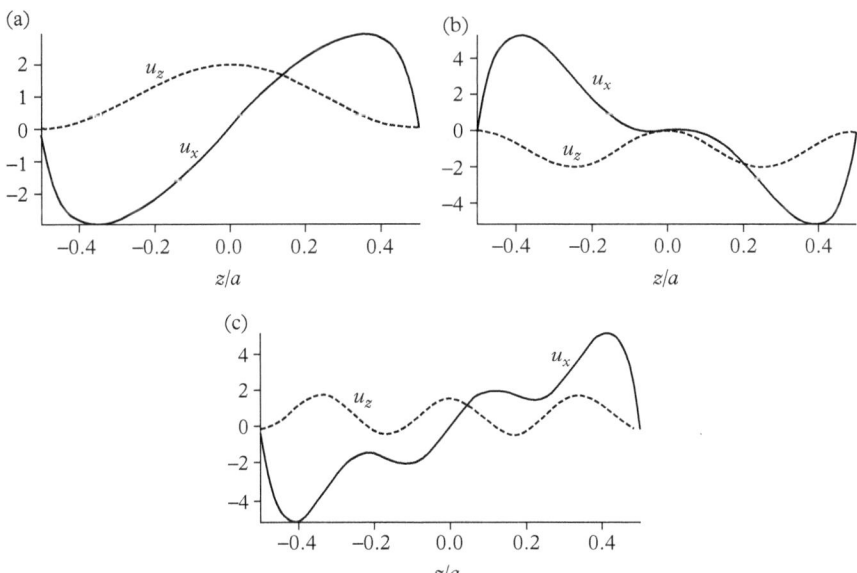

Figure 8.5 Patterns for IF-like modes for antisymmetric solutions $k_z L \sim$ (a) 2π, (b) 4π, (c) 6π. Reprinted with permission from Cambridge University Press.

88 Quantum Well

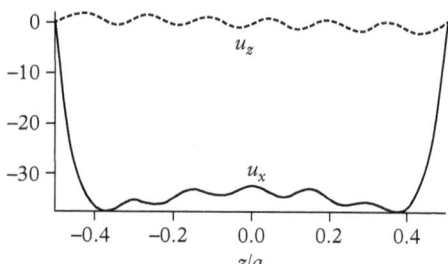

Figure 8.6 *A pattern of IF-like mode for a symmetric solution. Reprinted with permission from Cambridge University Press.*

8.2 Energy Normalization

Following the usual scheme for quantization for the triple hybrid, we obtain

$$A = \left(\frac{2}{K^2}\right)^{1/2} \left(\frac{\hbar}{2\mu\omega}\right)^{1/2} (a + a^\dagger) \tag{8.12}$$

For the symmetric solution:

$$K_s^2 = \begin{bmatrix} k_x^2 + k_z^2 + (k_x^2 - k_z^2)\frac{\sin k_z L}{k_z L} + 2\Gamma_{s1}^2 k_x^2 \frac{\sinh k_x L}{k_x L} - 4\Gamma_{s1} k_x^2 \frac{\sinh(k_x L/2)}{k_x L/2} \cos(k_z L/2) \\ +\Gamma_{s2}^2 \left(\frac{k_x^2}{\eta^2}\right)\left(k_x^2 - \eta^2 + (k_x^2 + \eta^2)\frac{\sin \eta L}{\eta L}\right) - 4\Gamma_{s2} k_x^2 \frac{\sinh(\eta L/2)}{(\eta L/2)} \cos(k_z L/2) \\ +4\Gamma_{s1}\Gamma_{s2} \frac{\cosh(\eta L/2)}{(\eta L/2)} \sinh(k_x L/2) \end{bmatrix} \tag{8.13}$$

For the antisymmetric solution:

$$K_a^2 = \begin{bmatrix} k_x^2 + k_z^2 - (k_x^2 - k_z^2)\frac{\sin k_z L}{k_z L} + 2\Gamma_{a1}^2 k_x^2 \frac{\sinh k_x L}{k_x L} - 4\Gamma_{a1} k_x^2 \frac{\sinh(k_x L/2)}{k_x L/2} \sin(k_z L/2) \\ +\Gamma_{a2}^2 \left(\frac{k_x^2}{\eta^2}\right)\left(k_x^2 - \eta^2 + (k_x^2 + \eta^2)\frac{\sin \eta L}{\eta L}\right) - 4\Gamma_{a2} k_x^2 \frac{\cosh(\eta L/2)}{(\eta L/2)} \sin(k_z L/2) \\ +4\Gamma_{a1}\Gamma_{a2} \frac{\sinh(\eta L/2)}{(\eta L/2)} \cosh(k_x L/2) \end{bmatrix} \tag{8.14}$$

Modes form double hybrids in non-polar quantum wells as they do in non-polar free-standing slabs. Hybrids in a multiwell structure in the form of a superlattice have been comprehensively studied and compared with computer-intensive lattice dynamical calculations, showing good agreement (Chamberlain et al. 1993). It is worthy of remark that a continuum theory can produce results that are in agreement with microscopic theory.

8.3 Reduced Boundary Condition

When the disparity between LO and TO frequencies is large enough ($\eta \to \infty$)—true for many cases in practice—the mechanical boundary condition simplifies to the requirement that only the normal component of the particle displacement need vanish; the TO components of the hybrid can be taken to automatically satisfy the condition that the tangential components vanish. Thus, in most cases, hybrid theory can reduce the number of boundary conditions from five to three: the continuity of the normal electric displacement, the continuity of the electric potential (or, equivalently, the continuity of the tangential components of the electric field), and the vanishing of the normal component of the particle displacement. Reduced hybrid theory exploits the very short spatial range of the evanescent TO component of the hybrid. It assumes that its effect via energy normalization on the interaction with, for example, electrons is negligible compared with that of the longer-range components. In short, it assumes that the role of the TO component is limited to satisfying the mechanical boundary condition on the tangential components of the displacement.

In the above equations, the effective elimination of the TO component is achieved by taking $\eta \to \infty$. The antisymmetric wave becomes

$$u_x = k_x A e^{ik_x x} [\sin k_z z - \Gamma_{a1} \sinh k_x z] (1 - \delta_{z,0})$$
$$u_z = ik_z A e^{ik_x x} [\cos k_z z - (k_x/k_z)\Gamma_{a1} \cosh k_x z] \qquad (8.15)$$

The symmetric wave is

$$u_x = k_x A e^{ik_x x} [\cos k_z z - \Gamma_{s1} \cosh k_x z] (1 - \delta_{z,0})$$
$$u_z = ik_z A e^{ik_x x} [\sin k_z z + (k_x/k_z)\Gamma_{s1} \sinh k_x z] \qquad (8.16)$$

The mechanical boundary condition is $u_z(\pm L/2) = 0$.

The energy normalization factors are, for the well, (for barrier see Section 8.5)

$$K_s^2 = \left[k_x^2 + k_z^2 + (k_x^2 - k_z^2) \frac{\sin k_z L}{k_z L} + 2\Gamma_{s1}^2 k_x^2 \frac{\sinh k_x L}{k_x L} - 4\Gamma_{s1} k_x^2 \frac{\sinh(k_x L/2)}{k_x L/2} \cos(k_z L/2) \right] \qquad (8.17)$$

$$K_a^2 = \left[k_x^2 + k_z^2 - (k_x^2 - k_z^2) \frac{\sin k_z L}{k_z L} + 2\Gamma_{a1}^2 k_x^2 \frac{\sinh k_x L}{k_x L} - 4\Gamma_{a1} k_x^2 \frac{\sinh(k_x L/2)}{k_x L/2} \sin(k_z L/2) \right] \qquad (8.18)$$

8.4 General Comments

An LO mode incident along the normal to the interface can satisfy all the connection rules without the aid of IF or TO modes. The tangential displacement is zero and, hence, so is the electric field; the normal particle displacement is finite but, because the dielectric function is zero, the electric displacement is also zero. The electric boundary conditions are, therefore, automatically satisfied. The mechanical boundary condition is satisfied by

$$k_z L/2 = n\pi \begin{cases} \text{symmetric } n \text{ even} \\ \text{antisymmetric } n \text{ odd} \end{cases} \qquad (8.19)$$

Once k_x is finite, a tangential electric field appears. The electric connection rules then require fields associated with the IF mode for their satisfaction. In the absence of lattice dispersion, the electric displacement introduced by the IF mode is zero, since the frequency of the IF mode must be the same as that of the LO mode, and its dielectric function is, therefore, zero. In that case, the satisfaction of the electrical boundary conditions also ensures that u_x is zero, so both electrical and mechanical boundary conditions can be satisfied without the introduction of the TO mode. Lattice dispersion makes the dielectric function of the IF mode non-zero, and it therefore necessitates the introduction of the TO mode in order to satisfy both electrical and mechanical boundary conditions. But, as pointed out in the previous section, its influence is heavily confined to the interface and its role limited, to a good approximation, in ensuring that u_x vanishes at the interface. The mechanical boundary condition reduces to $u_z = 0$ at the interface, and this can be ensured by the double hybrid LO/IF. As remarked earlier, the hybrid becomes IF-like over certain ranges of k_x.

These comments are also relevant to other nanostructures.

8.5 Barrier Modes

The barriers are taken to be very thick, which allows us to use the approach used in the case of the single heterojunction (SH). We need to describe the IF modes induced in the well at the barrier frequency.

Assuming the reduced mechanical boundary condition, $u_z = 0$ at the interface, we can write for the particle displacements in the barrier with $z \leq -L/2$:

$$u_x = k_x e^{ik_x x}(A \cos k_z z' + B \sin k_z z' + Ce^{k_x z'})(1 - \delta_{z',0})$$
$$u_z = ie^{ik_x x}(k_z A \sin k_z z' - k_z B \cos k_z z' - k_x Ce^{k_x z'}) \tag{8.20}$$

Here $z' = z + L/2$. The associated electric fields are

$$E_x = -\alpha_{0B} e^{ik_x x} k_x(A \cos k_z z' + B \sin k_z z' + s_B Ce^{k_x z'})$$
$$E_z = -i\alpha_{0B} e^{ik_x x}(k_z A \sin k_z z' - k_z B \cos k_z z' - s_B k_x Ce^{ik_x z'}) \tag{8.21}$$

The field in the well for the symmetric case is of the form

$$E_x^W = k_x F e^{ik_x x} z \cosh k_x z$$
$$E_z^W = -ik_x F e^{ik_x x} \sinh k_x \tag{8.22}$$

Satisfying the boundary conditions allows us to write

$$u_x = k_x e^{ik_x x} A(\cos k_z z' + \Gamma \sin k_z z' + p' e^{k_x z'})(1 - \delta_{z',0})$$
$$u_z = ie^{ik_x x} A(k_z \sin k_z z' - k_z \Gamma \cos k_z z' - k_x p' e^{k_x z'})$$
$$F = \frac{\alpha_{0B}(\varepsilon_B/\varepsilon_W) s_B p'}{\sinh(k_x L/2)} A \tag{8.23}$$

$$\Gamma = \frac{k_x}{k_z} p', \quad p' = \frac{1}{s_B[1 + (\varepsilon_B/\varepsilon_W)\coth(k_x L/2)]} \qquad (8.24)$$

The field induced in the well is associated with particle displacements that contribute to the mechanical energy:

$$\begin{aligned} u_x &= -\frac{1}{\alpha_{0W}} k_x F e^{ik_x x} \cosh k_x z \\ u_z &= \frac{i}{\alpha_{0W}} k_x F e^{ik_x x} \sinh k_x z \end{aligned} \qquad (8.25)$$

The energy normalization factor for the barrier modes becomes $K^2 = K_B^2 + K_W^2$. K_B^2 is given by equation (8.17). The addition due to the IF field induced in the well is

$$K_W^2 = \frac{k_x}{L} \left(\frac{\alpha_{0B}}{\alpha_{0W}}\right)^2 \frac{\coth(k_x L/2)}{[1 + (\varepsilon_B/\varepsilon_W)\coth(k_x L/2)]^2} \qquad (8.26)$$

A similar analysis can be made for the antisymmetric case.

8.6 Acoustic Modes

Acoustic hybrids, in what is essentially a double heterostructure (Wendler and Grigoryan 1988; Mitin et al. 1999), can be described following the methods used for the single heterostructure. Interface acoustic modes in a quantum well have been described by Wendler and Grigoryan (1988) and guided modes in acoustic wave guides, some properties of which are not dissimilar from those of a quantum well, by Auld (1990). In an isotropic solid, acoustic waves can be classified as transversely polarized (TA) or longitudinally polarized (LA). In real solids, this distinction can be made only in certain crystallographic directions of travel. Auld discusses the case of anisotropy; here, for simplicity, we will assume that the materials that make up the structure are elastically isotropic. Moreover, an isotropic material is always non-piezoelectric, simplifying matters further. A further classification of TA modes identifies waves as shear horizontal (SH) or shear vertical (SV). In a coordinate system in which the x direction represents the direction of unconfined travel, and the z direction is the direction normal to the interfaces or surfaces bounding the quantum well or slab, the components of SH waves are $(0, u_y, 0)$ and those of the SV waves are $(u_x, 0, u_z)$, which identifies them as what we have termed before as sTA and pTA modes respectively. The equivalent LA modes are pressure vertical (PV) or simply pLA modes.

The basic physics is encapsulated in the study of the reflection at and transmission through an interface, as we saw in the discussion of acoustic waves in single heterostructures in the last chapter—the sTA wave can satisfy the continuity of displacement and stress without hybridizing, whereas the pTA and pLA modes with the same frequency have to be linearly combined. In the case of the quantum well, however, the existence of two interfaces introduces confinement of modes that suffer total internal reflection. When the two interfaces are identical, as assumed here, the quantum well introduces two extra categories of wave patterns, symmetrical and antisymmetrical, as we saw in the case of optical modes.

8.6.1 sTA (Love Waves)

The quantum well can act as a leaky wave-guide, with sTA wave forms being a combination of confined and travelling waves:

$$u_y = e^{ik_x x}\left(A\begin{Bmatrix} \cos k_T z \\ \sin k_T z \end{Bmatrix} + B e^{ik_T z}\right) \tag{8.27}$$

corresponding to symmetric and antisymmetric patterns. In the barrier is a leaked travelling wave:

$$u_y = C e^{ik_x x} e^{iq_T z} \tag{8.28}$$

The relevant stress at the interface is

$$T_4 = c_{44}\frac{\partial u_y}{\partial z} e^{ik_x x} \tag{8.29}$$

(For brevity, we suppress the frequency factor.) Continuity of particle displacement and stress at the interface at $z = L/2$ gives two equations. With the mass approximation ($c_{44}^1 = c_{44}^2$), these become

$$A\begin{Bmatrix} \cos k_T L/2 \\ \sin k_T L/2 \end{Bmatrix} + B e^{ik_T L/2} = C e^{iq_t L/2}$$

$$A\begin{Bmatrix} -k_T \sin k_T L/2 \\ k_T \cos k_T L/2 \end{Bmatrix} + k_T B e^{ik_T z} = iq_T C e^{iq_T L/2} \tag{8.30}$$

Solving for the symmetric wave gives

$$k_T L/2 = n\pi,$$

$$B = -\frac{q_T}{q_T - k_T} A \cos n\pi, \quad C = -\frac{k_T}{q_T - k_T} A \cos n\pi \tag{8.31}$$

Closely related waves in a slab resting on a half-space, were described by A.E.H. Love in the context of seismology, and were termed leaky Love waves. In the case of the quantum well there is the possibility of propagating Love waves.

We consider the case in which the barriers have a smaller mass density, than the well, in which case $v_{T2} > v_{T1}$, where v_T is the velocity of the sTA wave. The frequencies have to be the same, so

$$q_T^2 = -k_x^2(1-\gamma_v) + \gamma_v k_T^2, \quad \gamma_v = v_{T1}^2/v_{T2}^2 \tag{8.32}$$

The condition for total internal reflection is $q_T = 0$, that is

$$k_x^2(1-\gamma_v) = \gamma_v k_T^2 \tag{8.33}$$

and $B = 0$. For k_x greater than this critical value, $q_T = i\eta_T$. The leaked wave in the barrier becomes an evanescent wave, and the particle displacements become

$$u_y = Ae^{ik_x x}\begin{cases}\cos k_T z \\ \sin k_T z\end{cases}\Big|_{-L/2 \geq z \leq L/2}, \quad u_y = Ce^{ik_x x}\begin{cases}e^{-\eta_T z} & z > L/2 \\ e^{\eta_T z} & z < -L/2\end{cases} \tag{8.34}$$

Satisfying the boundary conditions gives conditions on the wave vector. For the symmetrical solution:

$$\tan k_T L/2 = \eta_T/k_T \tag{8.35}$$

For the antisymmetrical solution:

$$\cot k_T L/2 = -\eta_T/k_T \tag{8.36}$$

The propagation velocity, v, defined by ω/k_x, is obtained from

$$\omega^2 = v_T^2(k_x^2 + k_T^2) = v^2 k_x^2 \tag{8.37}$$

For frequencies greater than $v_T k_x$, the wave vector is necessarily real. What about frequencies less than $v_T k_x$? Can the sTA wave become an interface mode? The answer is no, it cannot. For the sTA mode to become an interface mode, $k_T \to i\beta_T$, and equations (8.35) and (8.36) become

$$\begin{aligned}\tanh \beta_T L/2 &= -\frac{\eta_T}{\beta_T} \\ \coth \beta_T L/2 &= -\frac{\eta_T}{\beta_T}\end{aligned} \tag{8.38}$$

which cannot be satisfied.

8.6.2 p-modes

A pTA mode incident on an interface initiates transmitted pTA and pLA modes and reflected pTA and pLA modes. An incident pLA mode does the same. In a common case, the velocities in the well and barrier obey $v_{T1} < v_{L1} < v_{T2} < v_{L2}$, where the subscript 1 refers to the well and the subscript 2 refers to the barrier. The common frequency implies

$$\omega^2 = v_{T1}^2(k_x^2 + k_{T1}^2) = v_{L1}^2(k_x^2 + k_L^2) = v_{T2}^2(k_x^2 + k_{T2}^2) = v_{L2}^2(k_x^2 + k_{L2}^2) \quad (8.39)$$

Let us consider k_x to be fixed. It is clear that when $\omega < v_{T1}k_x$, all wave vectors must be imaginary to satisfy the common frequency condition. All p-modes are then interface modes (region I). New patterns emerge with increasing frequency. When $v_{T1}k_x < \omega < v_{L1}k_x$, the TA component becomes a guided wave in the well, and all other components retain their interface status (region II). When $v_{L1}k_x < \omega < v_{T2}k_x$, the LA component joins the TA component as a guided wave (region III). When $v_{T2}k_x < \omega < v_{L2}k_x$, the TA component becomes a leaky wave and the LA component remains a guided wave (region IV). When $v_{L2}k_x < \omega$, both components become leaky waves (region V).

Let us now consider the frequency as fixed. For $k_x < \omega/v_{L2}$, neither TA nor LA component is confined. When $k_x = \omega/v_{L2}$, the LA mode is totally internally reflected and becomes a guided wave; the TA component remains a leaky wave, and then becomes a guided wave when $k_x > \omega/v_{T2}$. Transitions to interface waves take place as $k_x > \omega/v_{L1}$ and $k_x > \omega/v_{T1}$.

Clearly, there are several families of modes, each occupying its region in $\omega - k_x$ space (Fig. 8.7). A particular component can be totally unconfined, partially confined, and completely confined either as a travelling wave or as an interface mode, dependent on the differences between the transverse and longitudinal velocities in the well and barrier. An extreme case is when the barrier density is much less than the well density, in which case $v_{T1} \ll v_{T2}$ and $v_{L1} \ll v_{L2}$. The modes are then the Lamb waves of the slab. When this is not the case, the rich complexity of patterns is known as generalized Lamb waves, or, for interface

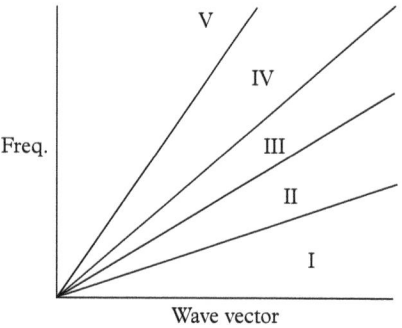

Figure 8.7 *Acoustic modes: frequency versus k_x. Regions for the case $v_{T1} < v_{L1} < v_{T2} < v_{L2}$. The lines represent the dispersion of the TA mode in the well, the LA mode in the well, the TA mode in the barrier, and the LA mode in the barrier. In region I the frequency is too low for the z component of the wave vectors to be real numbers; all modes are interface modes. In region II k_{T1} is real and the TA mode becomes a guided wave. In region III k_{L1} is real and the LA mode becomes a guided wave. In region IV k_{T2} is real and the TA guided wave becomes a leaky wave. In region V k_{L2} is real and both TA and LA modes are unconfined.*

modes, generalized Stonely waves. We limit subsequent discussion to interface modes and guided modes.

8.7 Interface Acoustic Waves

We focus on region I ($\omega < v_{T1}k_x$). In the well there will be symmetrical and antisymmetrical wave patterns. The components of the particle displacement for the symmetrical wave are

$$u_x = e^{ik_x x}(\beta_T A \cosh \beta_T z + k_x B \cosh \beta_L z)$$
$$u_z = e^{ik_x x}(-ik_x A \sinh \beta_T z - i\beta_L B \sinh \beta_L z) \tag{8.40}$$

(We have suppressed the frequency dependence and, as usual, introduced wave vector components to make obvious that $\nabla \cdot \mathbf{u} = 0$ for the TA mode and $\nabla \times \mathbf{u} = 0$ for the LA mode.) In the uppermost barrier layer:

$$u_x = e^{ik_x x}(\eta_T C e^{-\eta_L z} + k_x D e^{-\eta_L z})$$
$$u_z = e^{ik_x x}(ik_x C e^{-\eta_L z} + i\eta_L D e^{-\eta_L z}) \quad z \geq L/2 \tag{8.41}$$

In the lowermost barrier layer:

$$u_x = e^{ik_x x}(\eta_T C e^{\eta_L z} + k_x D e^{\eta_L z})$$
$$u_z = e^{ik_x x}(-ik_x C e^{\eta_L z} - i\eta_L D e^{\eta_L z}) \quad z \leq -L/2 \tag{8.42}$$

Satisfying the boundary conditions at $z = L/2$ gives four simultaneous equations for the amplitudes. Solutions exist provided the secular determinate vanishes. Thus

$$\begin{vmatrix} a_1 & b_1 & a_2 & b_2 \\ a'_1 & b'_1 & a'_2 & b'_2 \\ d_1 & e_1 & d_2 & e_2 \\ f_1 & g_1 & f_2 & g_2 \end{vmatrix} = 0 \tag{8.43}$$

$$\begin{bmatrix} a_1 = \beta_T \cosh \beta_T L/2 \\ a'_1 = k_x \sinh \beta_T L/2 \\ d_1 = -c^{(1)}_{44} 2k_x \beta_T \cosh \beta_T L/2 \\ f_1 = c^{(1)}_{44} (\beta_T^2 + k_x^2) \sinh \beta_T L/2 \end{bmatrix} \begin{bmatrix} b_1 = k_x \cosh \beta_L L/2 \\ b'_1 = \beta_L \sinh \beta_L L/2 \\ e_1 = c^{(1)}_{11}[k_x(1 - 2\gamma_1) - \beta_L^2] \cosh \beta_L L/2 \\ g_1 = c^{(1)}_{44} 2k_x \beta_L \sinh \beta_L L/2 \end{bmatrix}$$

$$\begin{bmatrix} a_2 = -\eta_T e^{-\eta_T L/2} \\ a'_2 = k_x e^{-\eta_T L/2} \\ d_2 = -c^{(1)}_{44} 2k_x \eta_T e^{-\eta_T L/2} \\ f_2 = c^{(1)}_{44} (\eta_T^2 + k_x^2) e^{-\eta_T L/2} \end{bmatrix} \begin{bmatrix} b_2 = -k_x e^{-\eta_L L/2} \\ b'_2 = \eta_L e^{-\eta_L L/2} \\ e_2 = c^{(2)}_{11}[k_x(1 - 2\gamma_1) - \beta_L^2]e^{-\eta_L L/2} \\ g_2 = c^{(2)}_{44} 2k_x \eta_L e^{-\eta_L L/2} \end{bmatrix}$$

$$\tag{8.44}$$

Here, c_{11} and c_{44} are the elastic constants. Along with the determinant are the relations associated with the common frequency:

$$\beta_T^2 = k_x^2(1 - s_1) \quad \eta_T^2 = k_x^2(1 - s_2)$$
$$\beta_L^2 = k_x^2(1 - \gamma_1 s) \quad \eta_L^2 = k_x^2(1 - \gamma_2 s_2) \tag{8.45}$$

$$s_1 = v/v_{T1} \quad s_2 = v/v_{T2} \quad v = \omega/k_x$$
$$\gamma_1 = c_{44}^{(1)}/c_{11}^{(1)} = v_{T1}/v_{L1} \quad \gamma_2 = c_{44}^{(2)}/c_{11}^{(2)} = v_{T2}/v_{L2} \tag{8.46}$$

A particularly simple case is when the barriers are replaced by vacuum, in which case the terms with subscript 2 vanish and the dispersion relation reduces to

$$-(2 - s_1)^2 \tanh \beta_T L/2 = 4(1 - s_1)^{1/2}(1 - \gamma_1 s_1)^{1/2} \tanh \beta_L L/2 = 0 \tag{8.47}$$

For the antisymmetric solution:

$$-(2 - s_1)^2 \tanh \beta_L L/2 = 4(1 - s_1)^{1/2}(1 - \gamma_1 s_1)^{1/2} \tanh \beta_T L/2 = 0 \tag{8.48}$$

For $L \to \infty$ we obtain the equation for Rayleigh waves. (Compare equations (6.21) and (6.23).)

Another simple case is when $k_x L \sim 0$. The dispersion relation reduces to

$$s_1 \beta_T \eta_L = 0 \tag{8.49}$$

The interface mode must vanish at $L = \infty$, so η_L cannot be zero; nor can s_1, so $\beta_T = 0$. This type of wave has been discussed by Sezawa (1927) for the case when there is one free surface. It is clear that we have here in the case of the quantum well a generalized Sezawa wave.

In general, however, the dispersion relation has to be solved numerically.

8.8 Guided Acoustic Waves

The above formalism can be applied to the case when both TA and LA components are guided waves (region III) by putting $\beta_T = ik_T$, $\beta_L = ik_L$. Replacing the barriers with the vacuum once again reduces the dispersion relation to that of the slab.

9
Quantum Wire

9.1 Introduction

Nanostructures in which the electrons are confined in two directions and unconfined in the third are known as quantum wires. Theories of electron confinement and phonon spectrum in quantum wires have focused on two simple geometries—wires with rectangular cross section and wires with circular cross section. As regards acoustic phonons, there is a comprehensive literature, as was mentioned in the Introduction. On the other hand, accounts of optical phonons have been limited to the results deriving from the dielectric continuum model, where it was noted that the corners inevitably associated with wires of rectangular cross section appeared to concentrate the interface mode (Knip and Reinecke 1992). Simple models applied to cylindrical waves were compared by Wang and Lei (1994), who, like Enderlein (1993), considered only independent LO and IF modes. Hybridization necessary to satisfy both electric and mechanical boundary conditions was not considered. Our focus here is to consider hybrid optical modes in a quantum wire and, for simplicity, we will consider the wire to have cylindrical symmetry surrounded by an infinitely thick cylindrical barrier, and we will continue to assume elastic anisotropy in the long-wavelength approximation, with the mechanical boundary condition for optical modes $\mathbf{u} = 0$.

The basic equation for the optical displacement $\mathbf{u}$ is

$$v_L^2 \nabla(\nabla \cdot \mathbf{u}) - v_T^2 \nabla \times \nabla \times \mathbf{u} = -(\omega_{L,T}^2 - \omega^2)\mathbf{u} \tag{9.1}$$

where v is the relevant velocity determining dispersion of the longitudinal and transverse waves. It is convenient to express the displacement in terms of a scalar potential, ϕ, and a vector potential, $\mathbf{A}$ (not to be confused with the corresponding electromagnetic potentials!), as follows:

$$\mathbf{u} = \nabla \phi + \nabla \times \mathbf{A} \tag{9.2}$$

Substitution into equation (9.1) gives

$$v_L^2 \nabla^2 \mathbf{u} - v_T^2 \nabla \times \nabla \times \nabla \times \mathbf{A} = -(\omega_{L,T}^2 - \omega^2)(\nabla \phi + \nabla \times \mathbf{A}) \tag{9.3}$$

Quantum Wire

We have used the identities $\nabla \cdot (\nabla \times \mathbf{A}) = 0, \nabla \times (\nabla \phi) = 0$. Another identity is $\nabla \times \nabla \times \mathbf{A} = \nabla(\nabla \cdot \mathbf{A}) - \nabla^2 \mathbf{A}$. We choose the vector potential to satisfy $\nabla \cdot \mathbf{A} = 0$, and equation (9.3) splits into separate equations for longitudinal and transverse modes, as follows:

$$v_L^2 \nabla^2 \phi = -(\omega_L^2 - \omega^2)\phi$$

$$v_T^2 \nabla^2 \mathbf{A} = -(\omega_T^2 - \omega^2)\mathbf{A} \qquad (9.4)$$

9.2 Cylindrical Coordinates

Equation (9.4) is truly independent of the coordinate system, but we now specialize to the coordinate system relevant to the cylinder. We take the z component to be along the length of the cylinder, with associated radial and angular components (Fig. 9.1). In cylindrical coordinates, the divergence of a gradient becomes the operator

$$\nabla^2 = \frac{\partial^2}{\partial r^2} + \frac{1}{r}\frac{\partial}{\partial r} + \frac{1}{r^2}\frac{\partial^2}{\partial \theta^2} + \frac{\partial^2}{\partial z^2} \qquad (9.5)$$

The ensuing calculation of the mode patterns of the optical phonons can be compared with the similar calculation for acoustic modes (Auld 1990).

9.2.1 Longitudinal Modes

The scalar potential for LO modes satisfies the equation

$$\nabla^2 \phi = \frac{\partial^2 \phi}{\partial r^2} + \frac{1}{r}\frac{\partial \phi}{\partial r} + \frac{1}{r^2}\frac{\partial^2 \phi}{\partial \theta^2} + \frac{\partial^2 \phi}{\partial z^2} = -\frac{\omega_L^2 - \omega^2}{v_L^2}\phi \qquad (9.6)$$

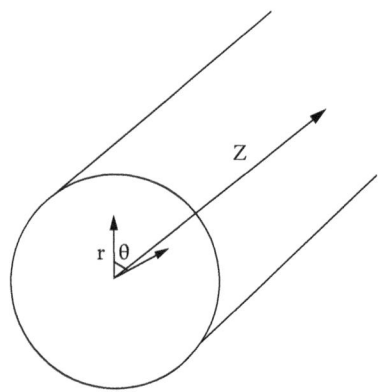

Figure 9.1 *The quantum wire.*

Expressing the potential according to $\phi = \phi_r(r)\phi_\theta(\theta)\phi_z(z)$, the equation becomes

$$\frac{1}{\phi_r}\left(\frac{\partial^2\phi_r}{\partial r^2} + \frac{1}{r}\frac{\partial\phi_r}{\partial r}\right) + \frac{1}{\phi_\theta r^2}\frac{\partial^2\phi_\theta}{\partial\theta^2} + \frac{1}{\phi_z}\frac{\partial^2\phi_z}{\partial z^2} = -\frac{\omega_L^2 - \omega^2}{v_L^2} \tag{9.7}$$

and we may take

$$\phi_\theta = e^{\pm im\theta}, \text{ alternatively } \phi_\theta = \begin{cases} \sin m\theta \\ \cos m\theta \end{cases}, \text{ also } \phi_z = e^{\pm ik_z z} \tag{9.8}$$

where $m = 0, \pm 1 \ldots \pm n$. The equation then becomes

$$\frac{1}{\phi_r}\left(\frac{\partial^2\phi_r}{\partial r^2} + \frac{1}{r}\frac{\partial\phi_r}{\partial r}\right) - \frac{m^2}{r^2} - k_z^2 = -\frac{\omega_L^2 - \omega^2}{v_L^2} \tag{9.9}$$

We now exploit the quadratic form of the dispersion relation by introducing the wave vector k_L such that

$$\omega^2 = \omega_L^2 - v_L^2(k_L^2 + k_z^2) \tag{9.10}$$

The radial component of the potential is then given by

$$\frac{1}{\phi_r}\left(\frac{\partial^2\phi_r}{\partial r^2} + \frac{1}{r}\frac{\partial\phi_r}{\partial r}\right) - \frac{m^2}{r^2} + k_L^2 = 0 \tag{9.11}$$

which is Bessel's equation, whose solution, finite at $r = 0$, is

$$\phi_r = \mathcal{J}_m(k_L r) \tag{9.12}$$

Here, $\mathcal{J}_m(\rho)$ is a Bessel function of the first kind. The lattice potential is then

$$\phi = A\mathcal{J}_m(k_L r)\begin{cases} \sin m\theta \\ \cos m\theta \end{cases} e^{ik_z z} \tag{9.13}$$

where A is a constant.

The components of the particle displacement, $\mathbf{u} = \nabla\phi$ (equation (9.2)), are

$$\begin{aligned} u_r &= \frac{\partial\phi}{\partial r} = Ak_L\mathcal{J}'_m(k_L r)\begin{cases} \sin m\theta \\ \cos m\theta \end{cases} e^{ik_z z} \quad \left(\mathcal{J}'_m(\rho) = \frac{\partial \mathcal{J}_m(\rho)}{\partial\rho}\right) \\ u_\theta &= \frac{1}{r}\frac{\partial\phi}{\partial\theta} = A\frac{m}{r}\mathcal{J}_m(k_L r)\begin{cases} \cos m\theta \\ -\sin m\theta \end{cases} e^{ik_z z} \\ u_z &= \frac{\partial\phi}{\partial z} = \pm iAk_z\mathcal{J}_m(k_L r)\begin{cases} \sin m\theta \\ \cos m\theta \end{cases} e^{ik_z z} \end{aligned} \tag{9.14}$$

100 Quantum Wire

We immediately see that the LO modes cannot satisfy the conditions $u_r = 0, u_\theta = 0$ at the interface. It is assumed that the wire is very long compared with its radius and that special boundary conditions at $z = 0$ and $z = L$ need not be considered.

9.2.2 Transverse Modes

The equation for the lattice vector potential is (equation (9.4))

$$v_T^2 \nabla^2 \mathbf{A} = -(\omega_T^2 - \omega^2)\mathbf{A} \qquad (9.15)$$

Quadratic dispersion is assumed as usual, and it is useful to define the quantity ψ via $\mathbf{A} = \psi \mathbf{z}$, as noted in equations (5.8) and (5.9):

$$\omega^2 = \omega_T^2 - v_T^2(k_T^2 + k_z^2)$$

$$\nabla^2 \psi = -\left(\frac{\omega_T^2 - \omega^2}{v_T^2}\right)\psi$$

$$\psi = \mathcal{J}_m(k_T r) \begin{Bmatrix} \sin m\theta \\ \cos m\theta \end{Bmatrix} e^{ik_z z} \qquad (9.16)$$

Note that ψ here is a scalar quantity.

It is useful to recall that the divergence and curl in cylindrical coordinates are

$$\nabla \cdot \mathbf{A} = \frac{\partial A_r}{\partial r} + \frac{A_r}{r} + \frac{1}{r}\frac{\partial A_\theta}{\partial \theta} + \frac{\partial A_z}{\partial z}$$

$$\nabla \times \mathbf{A} = \left(\frac{1}{r}\frac{\partial A_z}{\partial \theta} - \frac{\partial A_\theta}{\partial z}\right)\hat{\mathbf{r}} + \left(\frac{\partial A_r}{\partial z} - \frac{\partial A_z}{\partial r}\right)\hat{\boldsymbol{\theta}} + \frac{1}{r}\left(\frac{\partial(rA_\theta)}{\partial r} - \frac{\partial A_r}{\partial \theta}\right)\hat{\mathbf{z}} \qquad (9.17)$$

As observed in equations (5.9) and (5.10), there are two distinct solutions for the TO modes: $\mathbf{A}_1 = \nabla \times \psi(\hat{\mathbf{z}})$, $\mathbf{A}_2 = (1/k)\nabla \times \mathbf{A}_1$.

The components are

$$\mathbf{A}_1 = C_1 e^{ik_z z}\left[\frac{m}{r}\mathcal{J}_m(k_T r)\begin{Bmatrix} c \\ -s \end{Bmatrix}\hat{\mathbf{r}} - k_T \mathcal{J}'_m(k_T r)\begin{Bmatrix} s \\ c \end{Bmatrix}\hat{\boldsymbol{\theta}}\right] \qquad (9.18a)$$

$$\mathbf{A}_2 = C_2 e^{ik_z z}\left[ik_z k_T \mathcal{J}'_m(k_T r)\begin{Bmatrix} s \\ c \end{Bmatrix}\hat{\mathbf{r}} + ik_z \frac{m}{r}\mathcal{J}_m(k_T r)\begin{Bmatrix} c \\ -s \end{Bmatrix}\hat{\boldsymbol{\theta}} - k_T^2 \mathcal{J}_m(k_T r)\begin{Bmatrix} s \\ c \end{Bmatrix}\hat{\mathbf{z}}\right]$$

$$(9.18b)$$

The associated particle displacements for TO($\mathbf{A}_1$) are

$$\mathbf{u}_1 = \nabla \times \mathbf{A}_1 = C_1 e^{ik_z z}\left[ik_z k_T \mathcal{J}'_m(k_T r)\begin{Bmatrix} s \\ c \end{Bmatrix}\hat{\mathbf{r}} + i\frac{k_z m}{r}\mathcal{J}_m(k_T r)\begin{Bmatrix} c \\ -s \end{Bmatrix}\hat{\boldsymbol{\theta}} - k_T^2 \mathcal{J}_m(k_T r)\begin{Bmatrix} s \\ c \end{Bmatrix}\hat{\mathbf{z}}\right]$$

$$(9.19)$$

For TO(A_2) they are

$$\mathbf{u}_2 = \nabla \times \mathbf{A}_2 = C_2 e^{ik_z z}\left[-(k_T^2 - k_z^2)\frac{m}{r}\mathcal{J}_m(k_T r)\begin{Bmatrix}c\\-s\end{Bmatrix}\hat{\mathbf{r}} + k_T(k_T^2 - k_z^2)\mathcal{J}'_m(k_T r)\begin{Bmatrix}s\\c\end{Bmatrix}\hat{\boldsymbol{\theta}}\right] \quad (9.20)$$

where C_1 and C_2 are amplitudes. (Note that TO(A_2) has no z component.) An obvious notation, $s = \sin m\theta$, $c = \cos m\theta$, in the brackets has been used. The TO(A_2) mode, having no component in the z direction, can be regarded as an s-wave, the TO(A_1) as a p-wave.

Equation (9.14) gives the displacement components for the LO mode and equations (9.19) and (9.20) give the displacement components for the TO modes. We now have to consider the effects of the boundary conditions.

9.3 Interface Modes

In the case of non-polar and polar wires, the relevant mechanical boundary condition is $\mathbf{u} = 0$ at the surface $r = R$, for all modes. In polar material, there are the additional boundary conditions associated with the electric fields of the LO modes which entail the existence of an electric interface mode—essentially, an interface polariton far from the light line. Such a mode must have a frequency equal to that of the LO mode. Being an electromagnetic wave, it is transversely polarized with a frequency determined by the velocity of light, c, in the medium:

$$\omega^2 = c^2(k_z^2 + k_{IF}^2) = v_L^2(k_z^2 + k_L^2)$$
$$c^2 \gg v_L^2$$
$$k_{IF}^2 \approx -k_z^2 \quad (9.21)$$

Solving Laplace's equation for the electric potential gives

$$\Phi_E = B_1 I_m(k_z r)\begin{Bmatrix}s\\c\end{Bmatrix} e^{ik_z z} \quad (9.22)$$

where I_m is the modified Bessel function, finite at $r = 0$. External to the cylinder, the potential is

$$\Phi_E = B_2 K_m(k_z r)\begin{Bmatrix}s\\c\end{Bmatrix} e^{ik_z z} \quad (9.23)$$

where K_m is the modified Bessel function, zero at infinity. The electric fields are obtained from $\mathbf{E} = -\nabla\Phi_E$, so the LO displacement and the IF fields are given by

$$\mathbf{u_L} = A\left(k_L \mathcal{J}'_m(k_L r)\Big|^s_c \hat{\mathbf{r}} + \frac{m}{r}\mathcal{J}_m(k_L r)\Big|^c_{-s} \hat{\theta} + ik_z\mathcal{J}_m(k_L r)\Big|^s_c\right)e^{ik_z z} \quad (9.24)$$

$$\mathbf{E_1} = -B_1\left(k_z I'_m(k_z r)\Big|^s_c \hat{\mathbf{r}} + \frac{m}{r}I_m(k_z r)\Big|^c_{-s} \hat{\theta} + ik_z I_m(k_z r)\Big|^s_c \hat{\mathbf{z}}\right)e^{ik_z z} \quad r \leq R \quad (9.25)$$

$$\mathbf{E_2} = -B_2\left(k_z K'_m(k_z r)\Big|^s_c \hat{\mathbf{r}} + \frac{m}{r}K_m(k_z r)\Big|^c_{-s} \hat{\theta} + ik_z K_m(k_z r)\Big|^s_c \hat{\mathbf{z}}\right)e^{ik_z z} \quad r \geq R \quad (9.26)$$

As mentioned in Chapter 8, we need consider only the IF mode at the well frequency in the barrier.

The condition at $r = R$ is that the normal electric displacement and tangential fields are continuous. Noting that the LO displacement is zero leads to

$$\hat{\mathbf{r}} \quad B_1 \varepsilon_{1IF}(\omega) I'_m(k_z R) = B_2 \varepsilon_{2IF} K'_m(k_z R)$$

$$\hat{\theta} \text{ and } \hat{\mathbf{z}} \quad A\alpha_0 \mathcal{J}_m(k_L R) + B_1 I_m(k_z R) = B_2 K_m(k_z R)$$

$$\varepsilon(\omega_{SO}) = \varepsilon_\infty \frac{\omega_{IF}^2 - \omega_L^2}{\omega_{IF}^2 - \omega_T^2} \left\{\omega_{IF}^2 = \omega_L^2 - v_L^2(k_z^2 + k_L^2)\right\} \quad (9.27)$$

where A, B_1, and B_2 are the relevant amplitudes. Also, $\alpha_0 = e^*/V_0 \varepsilon_\infty$ Thus

$$B_1 = A \frac{\alpha_0 \mathcal{J}_m(k_L R) K'_m(k_z R)}{I'_m(k_z R) K_m(k_z R)} \left(\left[p(\omega_{SO}) - \frac{I_m(k_z R) K'_m(k_z R)}{I'_m(k_z R) K_m(k_z R)}\right]\right)^{-1}$$

$$p(\omega_{SO}) = \varepsilon_1(\omega)/\varepsilon_2(\omega) \quad (9.28)$$

Subscript 1 refers to the well ($z \leq R$), whereas subscript 2 to the barrier ($z \geq R$). (Note that we cannot use 'r' for the permittivity ratio; we use p instead.) Elimination of the z component of particle displacement is effected by TO(A_1), elimination of the θ component by TO(A_2). In the case of cylindrical coordinates both TO modes are necessary.

Related to this field are the optical mode displacements:

$$-\alpha_0 s_1(\omega_{IF})\mathbf{u}_{1IF} = \mathbf{E_1} \text{ in well}, \quad -\alpha_0 s_2 \mathbf{u}_{2IF} = \mathbf{E_2} \text{ in barrier} \quad (9.29)$$

Here, s is the field factor. For the LO mode the field factor is always unity, but for the IF mode it is less than unity as a result of dispersion.

9.4 Hybrid Modes in Polar Material

We focus attention on LO modes in polar material. We notice that the LO mode cannot satisfy the mechanical boundary condition, $\mathbf{u} = 0$, on its own, nor can it satisfy the electrical boundary conditions without forming a hybrid with the IF mode. We therefore consider the linear combination LO + IF + TO(A_1) + TO(A_2), each sharing a common frequency, namely, that of the LO mode. This means that the TO components have to be evanescent modes with large imaginary wave vector, η. The particle displacements at $r = R$ are

$$\hat{\mathbf{r}} \quad \left. \left(A(k_L\mathcal{J}'_m + \Gamma k_z I'_m) - C_1(i)^m \eta k_z I'_{mT}\right)\right|^s_c + \left. C_2(i)^m \frac{m}{R}(-\eta^2 + k_z^2) I_{mT}\right|^{-s}_c = 0$$

$$\hat{\boldsymbol{\theta}} \quad \left. \frac{m}{R}\left(A(\mathcal{J}_m + \Gamma I_m) + C_1(i)^{m+1} \eta I_{mT}\right)\right|^{-s}_c - \left. C_2(i)^{m+1} \eta(\eta^2 + k_z^2) I'_{mT}\right|^c_s = 0$$

$$\hat{\mathbf{z}} \quad \left. \left(ik_z A(\mathcal{J}_m + \Gamma I_m) + C_1(i)^m \eta^2 I_{mT}\right)\right|^s_c = 0$$

$$\Gamma = \frac{\mathcal{J}_m K'_m}{I'_m K_m} \left[s(\omega_{IF}) \left(p(\omega_{IF}) - \frac{I_m K'_m}{I'_m K_m} \right) \right]^{-1} \tag{9.30}$$

and s is the field factor. (The arguments of the Bessel functions have been suppressed for brevity. For $\mathcal{J}$ the argument is $k_L R$, for I_m it is $k_z R$ and for I_{mT} is ηR. Also, the frequency factor has been suppressed.) All amplitudes can be expressed in terms of A, and A is obtained by the usual process of energy normalization (see following).

Solutions for the three amplitudes exist, provided that the determinant of the coefficients vanishes. A simple result for the dispersion is obtained in the limit $\eta \to \infty$, namely:

$$k_L \mathcal{J}'_m(k_L r) + \Gamma k_z I'_m(k_z r) = 0 \tag{9.31}$$

This limit provides the basis for the double-hybrid theory, in which the particle displacements of the polar mode are

$$\mathbf{u} = A e^{i(k_z z - \omega t)} \begin{pmatrix} [k_L \mathcal{J}'_m(k_L r) + \Gamma k_z I'_m(k_z r)] \left|\begin{matrix}\sin m\theta \\ \cos m\theta\end{matrix}\right. \hat{\mathbf{r}} \\ + \frac{m}{r} [\mathcal{J}_m(k_L r) + \Gamma I_m(k_z r)] \left|\begin{matrix}-\sin m\theta \\ \cos m\theta\end{matrix}\right. \hat{\boldsymbol{\theta}} \\ + ik_z [\mathcal{J}_m(k_L r) + \Gamma I_m(k_z r)] \left|\begin{matrix}\sin m\theta \\ \cos m\theta\end{matrix}\right. \hat{\mathbf{z}} \end{pmatrix} \tag{9.32}$$

$$\Gamma = \frac{\mathcal{J}_m(k_L R) K'_m(k_z R)}{I'_m(k_z R) K_m(k_z R)} \frac{1}{s\left(p - \frac{I_m(k_z R) K'_m(k_z R)}{I'_m(k_z R) K_m(k_z R)}\right)} \tag{9.33}$$

104 Quantum Wire

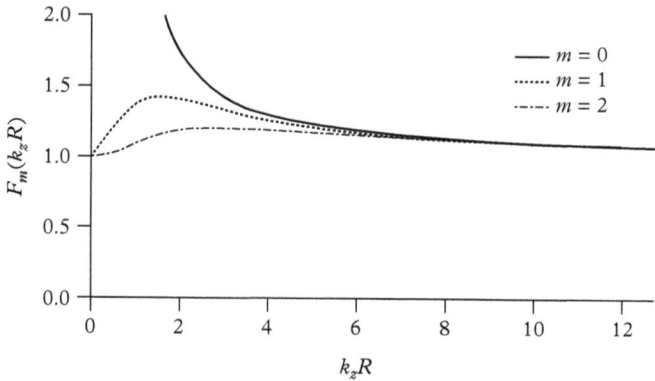

Figure 9.2 *The dependence of the function F_m on $k_z R$.*

It is useful to first consider the pattern of dispersion for $k_z = 0$. In this case, the solutions for $k_L R$ are the zeros of $\mathcal{J}'_m(k_L R)$. As k_z increases, the resonant condition is reached when $p + F_m = 0$, where

$$F_m = -\frac{I_m(k_z R) K'_m(k_z R)}{I'_m(k_z R) K_m(k_z R)}$$

Its dependence on k_z is depicted in Figure 9.2.

It is clear that there is a significant difference between modes with $m = 0$ and $m \neq 0$. For $m = 0$, F_m is very large at modest k_z, and so avoids the resonant condition. Like the symmetrical modes of the quantum well, the $m = 0$ modes are essentially LO-like. Their dispersion is shown in Figure 9.3 for a freestanding GaAs quantum wire of radius 5 nm. For $m \neq 0$, F_m has a magnitude

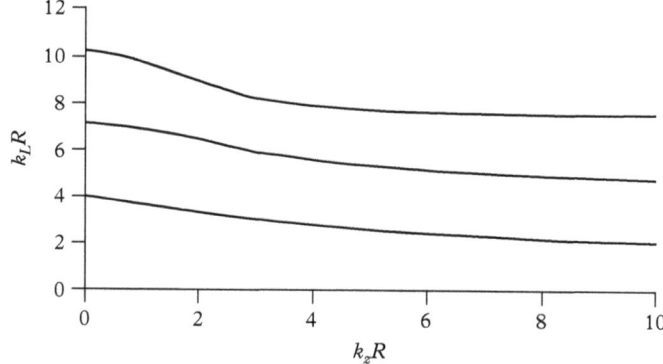

Figure 9.3 *Quantum wire dispersion for $m = 0$: $R = 5$ nm.*

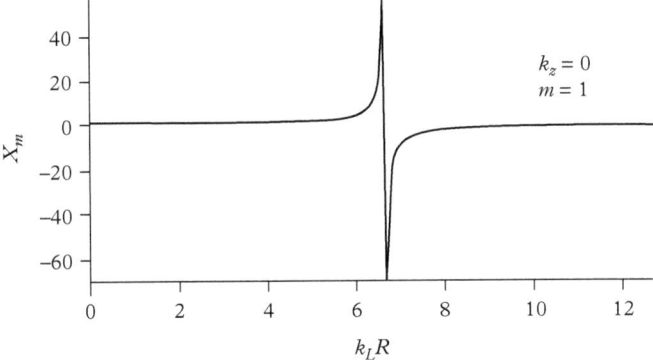

Figure 9.4 Dependence of the SO function X_m on $k_L R$ for $k_z = 0$.

near to unity, varying very little with k_z. The resonance condition therefore requires the permittivity factor to be about −1, which will occur at a particular value of $k_L R$, determined by the lattice dispersion. This is illustrated in Figure 9.4 by the dependence on $k_L R$ of the factor:

$$X_m = \frac{F_m}{s(p + F_m)}$$

The dispersion for $m = 1$ is shown in Figure 9.5 for a free-standing GaAs quantum wire of radius 5 nm.

Energy normalization (quantization) gives for the amplitude (cf. equation (8.12)):

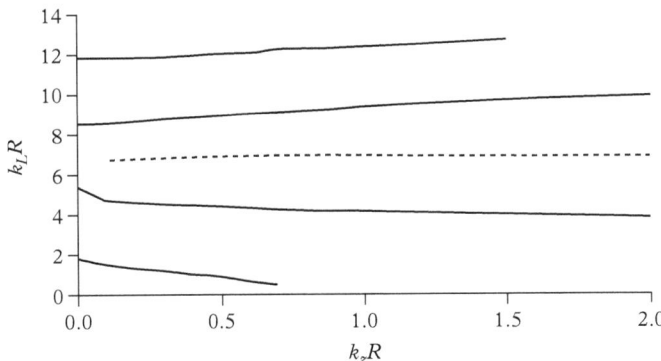

Figure 9.5 Quantum wire dispersion for $m = 1$: $R = 5$ nm. The dashed curve depicts the dispersion of the SO-like hybrid.

Quantum Wire

$$A = \left(\frac{2}{K^2}\right)^{1/2} \left(\frac{\hbar}{2\bar{M}\omega}\right)^{1/2} (a + a^\dagger)$$

$$K^2/2 = \int \mathbf{u}^* \cdot \mathbf{u} \, dr / (\pi R^2 L_z) \tag{9.34}$$

For $m = 0$,

$$K_0^2 = K_{0r}^2 + K_{0\theta}^2 + K_{0z}^2 \tag{9.35}$$

$$K_{0r}^2 = \left(k_L^2 (\mathcal{J}_1^2 - \mathcal{J}_0 \mathcal{J}_2) - k_z^2 \Gamma^2 (I_1^2 - I_0 I_2) - \frac{4 k_L k_z}{(k_L^2 + k_z^2) R} \Gamma(k_z \mathcal{J}_1 I_0 + k_L \mathcal{J}_0 I_1) \right)$$

$$K_{0\theta}^2 = 0$$

$$K_{0z}^2 = k_z^2 \left((\mathcal{J}_0^2 + \mathcal{J}_1^2) + \Gamma^2 (I_0^2 - I_1^2) + \frac{4}{(k_L^2 + k_z^2) R} (k_z \mathcal{J}_0 I_1 - k_L \mathcal{J}_1 I_0) \right) \tag{9.36}$$

This reduces to the DC result for the LO mode (Wang and Lei 1994) with $\mathcal{J}_0 = 0$. The contribution to the energy in the barrier is negligible.

9.5 Acoustic Stresses and Strains

In cubic material, the six strain components, in cylindrical coordinates, are

$$S_{rr} = S_1 = \frac{\partial u_r}{\partial r}$$

$$S_{\theta\theta} = S_2 = \frac{\partial u_\theta}{r \partial \theta} + \frac{u_r}{r}$$

$$S_{zz} = S_3 = \frac{\partial u_z}{\partial z}$$

$$S_{\theta z} = S_4 = \frac{\partial u_\theta}{\partial z} + \frac{\partial u_z}{r \partial \theta}$$

$$S_{zr} = S_5 = \frac{\partial u_z}{\partial r} + \frac{\partial u_r}{\partial z}$$

$$S_{r\theta} = S_6 = \frac{\partial u_r}{r \partial \theta} + \frac{u_\theta}{r} + \frac{\partial u_\theta}{\partial r} \tag{9.37}$$

The corresponding stresses are

$$T_{rr} = T_1 = c_{11}S_1 + c_{12}(S_2 + S_3)$$

$$T_{\theta\theta} = T_2 = c_{11}S_2 + c_{12}(S_3 + S_1)$$

$$T_{zz} = T_3 = c_{11}S_3 + c_{12}(S_1 + S_2)$$

$$T_{\theta z} = T_4 = c_{44}S_4$$

$$T_{zr} = T_5 = c_{44}S_5$$

$$T_{r\theta} = T_6 = c_{44}S_6 \tag{9.38}$$

The particle displacements in the well are obtained from the lattice scalar and vector potentials of Chapter 5:

$$LA \quad \mathbf{u} = Ae^{ik_z z}\left[k_L \mathcal{J}'_m(k_L r)\begin{Bmatrix}s\\c\end{Bmatrix}\hat{\mathbf{r}} + \frac{m}{r}\mathcal{J}_m(k_L r)\begin{Bmatrix}c\\-s\end{Bmatrix}\hat{\boldsymbol{\theta}} + ik_z\mathcal{J}_m(k_L r)\begin{Bmatrix}s\\c\end{Bmatrix}\hat{\mathbf{z}}\right]$$

$$TA_1 \quad \mathbf{u} = Be^{ik_z z}\left[ik_z k_T \mathcal{J}'_m(k_T r)\begin{Bmatrix}s\\c\end{Bmatrix}\hat{\mathbf{r}} + ik_z\frac{m}{r}\mathcal{J}_m(k_T r)\begin{Bmatrix}c\\-s\end{Bmatrix}\hat{\boldsymbol{\theta}} - k_T^2 \mathcal{J}_m(k_L r)\begin{Bmatrix}s\\c\end{Bmatrix}\hat{\mathbf{z}}\right]$$

$$TA_2 \quad \mathbf{u} = Ce^{ik_z z}\left[-\frac{m}{r}\mathcal{J}_m(k_T r)\begin{Bmatrix}c\\-s\end{Bmatrix}\hat{\mathbf{r}} + k_T \mathcal{J}'_m(k_T r)\begin{Bmatrix}s\\c\end{Bmatrix}\hat{\boldsymbol{\theta}}\right] \tag{9.39}$$

In the barrier they are expressed in terms of associated Bessel functions that vanish at $r = \infty$:

$$LA \quad \mathbf{u} = De^{ik_z z}\left[\eta_L K'_m(\eta_L r)\begin{Bmatrix}s\\c\end{Bmatrix}\hat{\mathbf{r}} + \frac{m}{r}K_m(\eta_L r)\begin{Bmatrix}c\\-s\end{Bmatrix}\hat{\boldsymbol{\theta}} + ik_z K_m(\eta_L r)\begin{Bmatrix}s\\c\end{Bmatrix}\hat{\mathbf{z}}\right]$$

$$TA_1 \quad \mathbf{u} = Ee^{ik_z z}\left[-k_z\eta_T K'_m(\eta_T r)\begin{Bmatrix}s\\c\end{Bmatrix}\hat{\mathbf{r}} + ik_z\frac{m}{r}K_m(\eta_T r)\begin{Bmatrix}c\\-s\end{Bmatrix}\hat{\boldsymbol{\theta}} + \eta_T^2 K_m(\eta_T r)\begin{Bmatrix}s\\c\end{Bmatrix}\hat{\mathbf{z}}\right]$$

$$TA_2 \quad \mathbf{u} = Fe^{ik_z z}\left[\frac{m}{r}K_m(\eta_T r)\begin{Bmatrix}c\\-s\end{Bmatrix}\hat{\mathbf{r}} - \eta_T K'_m(\eta_T r)\begin{Bmatrix}s\\c\end{Bmatrix}\hat{\boldsymbol{\theta}}\right] \tag{9.40}$$

Notice that the TA_2 modes lack a z component. In some of these derivations, it is sometimes useful to recall that

$$\mathcal{J}''_m + \frac{\mathcal{J}'_m}{kr} + \left(1 - \frac{m^2}{(kr)^2}\right)\mathcal{J} = 0 \tag{9.41}$$

Satisfying the boundary conditions—the continuity of displacement and of stress—generates many families of modes, as it did in the case of the quantum well: unconfined modes, leaky modes, guided modes, and interface modes, in their respective regions in ω, k_z space. A general account would need a book in its own right; here we limit attention to the simplest cases.

9.6 Free Surface

We focus on elastically isotopic quantum wires with a free surface. In this case the stress components must vanish at $r = R$. We further limit attention to the symmetrical azimuthal case, that is, $m = 0$. Solutions for A, B, and C exist provided that the characteristic determinant vanishes:

$$\left(k_T^2 \mathcal{J}_0(k_T R) + \frac{2k_T}{R}\mathcal{J}_0'(k_T R)\right) \times Det = 0$$

$$Det = \begin{vmatrix} 2c_{44}[k_L^2(\mathcal{J}_0''(k_L R) + k_z^2 \mathcal{J}_0(k_L R)] - c_{11}(k_L^2 + k_z^2)\mathcal{J}_0(k_L R) & 2c_{44}ik_T k_z \mathcal{J}_0''(k_T R) \\ 2ik_L k_z \mathcal{J}_0'(k_L R) & (k_T^2 - k_x^2)\mathcal{J}_0(k_T R) \end{vmatrix}$$

(9.42)

The LA and TA$_1$ mode remains coupled, but the TA$_2$ mode is completely decoupled. In this respect, and its ability to satisfy the stress condition $T_6 = 0$ without assistance, makes the TA$_2$ mode the cylindrical doppelganger of the sTA (SH) mode in Cartesian coordinates. These symmetrical azimuthal modes have been termed torsional modes, in which the allowed wave vectors are determined by

$$k_T^2 \mathcal{J}_0(k_T R) + \frac{2k_T}{R}\mathcal{J}_0'(k_T R) = 0 \tag{9.43}$$

and $u_r = u_z = 0$.

The LA and TA$_1$ modes decouple in the special case $k_z = 0$. The LA modes are purely radial, with $u_\theta = u_z = 0$, and k_L determined by

$$2c_{44}k_L^2\{\mathcal{J}_0''(k_L R) + \mathcal{J}_0(k_L R)\} - k_L^2 c_{11}\mathcal{J}_0(k_L R) = 0 \tag{9.44}$$

The TA$_1$ modes are purely axial, with $u_r = u_\theta = 0$, and

$$\mathcal{J}_0(k_T R) = 0 \tag{9.45}$$

These decoupled modes are termed dilatational.

When $k_z \neq 0$, the modes are coupled and in the low frequency region (region I) they become interface modes, or, in the case of a free surface, surface modes.

The transition from real wave vectors to imaginary vectors when $m = 0$ means $k_L \to i\beta_L$, $k_T \to i\beta_T$, $\mathcal{J}_0 \to I_0$, $\mathcal{J}_0' \to -iI_0'$, $\mathcal{J}_0'' \to -I_0''$, where I_0 is the associated Bessel function obeying

$$I_0'' + \frac{I_0'}{\beta r} - I_0 = 0 \tag{9.46}$$

We put $\omega^2 = v^2 k_z^2$, $s = v^2/v_T^2$, and $\beta_T^2 = k_z^2(1-s)$, $\beta_L^2 = k_z^2(1-\gamma s)$, $\gamma = c_{44}/c_{11} = v_T^2/v_L^2$. The characteristic determinant becomes

$$\{2I_0''(\beta_L R)(1-\gamma s) - s(1-2\gamma)I_0(\beta_L R)\}(2-s)I_0'(\beta_T R)$$
$$- 4(1-s)^{1/2}(1-\gamma s)^{1/2}I_0'(\beta_L R)I_0''(\beta_T R) = 0 \tag{9.47}$$

In the limit $\beta R \to \infty$, this reduces to the familiar Rayleigh equation for surface waves:

$$(2-s)^2 - 4(1-s)^{1/2}(1-\gamma s)^{1/2} = 0 \tag{9.48}$$

In the general case, there will exist a family of modes for each value of m. The total number of modes is, of course, limited by the dimensions of the crystal.

10
Quantum Dot

10.1 Introduction

Interest in the physics of quantum dots was stimulated by the existence of colourful glasses and the possibility of fabricating macroscopic analogues of atoms. Silicon dioxide is colourless, so to get coloured glass it was necessary to incorporate a suitable substance that absorbed light at characteristic wavelengths in the glass melt. As the glass cooled, the dilute additive material crystallized into tiny crystallites, spread throughout the glass. Modern techniques involving the temperature and heat treatment of the glass can, to some degree, obtain crystallites of uniform size, and these turn out in dilute glasses to be roughly spherical with nanometre radii. Other substances that exhibit quantum dots are colloids, in which glass is replaced by a liquid solvent. Colloids have the advantage that they allow manipulation of the quantum dot population. In both techniques, the dopant is usually a semiconductor. Other techniques, such as ion implantation, are also used (e.g. Banyai and Koch 1993).

Experimental determination of the properties of quantum dots is bedevilled by the spread in sizes of dots in the population. Nevertheless, interest in their optical properties and their dependence on the degree of confinement of electrons, holes, and lattice waves has stimulated a considerable literature. In particular, the excitation of electrons and holes, the nature of the exciton, the interaction with phonons, and the anharmonic effects associated with the Stokes shift have received considerable attention. The common approach to the electron–phonon interaction has been via the DC model. Here, we focus attention on the confinement of lattice waves, in particular, of polar optical modes and their hybridization.

For simplicity, we will consider the dot to be a sphere of an elastically isotropic semiconductor.

10.2 Spherical Coordinates

The particle displacement for the LO mode is expressed in terms of a scalar potential, Φ, as described in Chapter 5. The Helmholtz equation that describes the scalar potential in spherical polar coordinates is

$$\frac{1}{r^2}\frac{\partial}{\partial r}\left(r^2\frac{\partial\Phi}{\partial r}\right) + \frac{1}{r^2\sin\theta}\frac{\partial}{\partial\theta}\left(\sin\theta\frac{\partial\Phi}{\partial\theta}\right) + \frac{1}{r^2\sin^2\theta}\frac{\partial^2\Phi}{\partial\phi^2} + k^2\Phi = 0 \qquad (10.1)$$

This has the solution

$$\Phi = j_\ell(kr)P_\ell^m(\theta)\begin{cases}\cos m\phi \\ \sin m\phi\end{cases} \qquad (10.2)$$

Here $j_\ell(\rho)$ is a spherical Bessel function and $P_\ell^m(\theta)$ is an associated Legendre polynomial; ℓ is an integer and $-\ell \leq m \leq \ell$. We note that

$$\nabla\Phi = \frac{\partial\Phi}{\partial r}\hat{\mathbf{r}} + \frac{1}{r}\frac{\partial\Phi}{\partial\theta}\hat{\boldsymbol{\theta}} + \frac{1}{r\sin\theta}\frac{\partial\Phi}{\partial\phi}\hat{\boldsymbol{\phi}}$$

$$\nabla\times\mathbf{A} = \frac{1}{r\sin\theta}\left[\frac{\partial}{\partial\theta}(A_\phi\sin\theta) - \frac{\partial A_\theta}{\partial\phi}\right]\hat{\mathbf{r}} + \left[\frac{1}{r\sin\theta}\frac{\partial A_r}{\partial\phi} - \frac{1}{r}\frac{\partial}{\partial r}(rA_\phi)\right]\hat{\boldsymbol{\theta}} \qquad (10.3)$$

$$+ \frac{1}{r}\left[\frac{\partial}{\partial r}(rA_\theta) - \frac{\partial A_r}{\partial\theta}\right]$$

10.2.1 Polar Modes

The components of the particle displacement for LO modes are derived from the lattice scalar potential: $\mathbf{u} = \nabla\phi$:

$$\mathbf{u} = A\begin{pmatrix}kj'_\ell(kr)P_\ell^m(\theta)\begin{cases}\cos m\phi \\ \sin m\phi\end{cases}\hat{\mathbf{r}} + \frac{1}{r}j_\ell(kr)\frac{\partial P_\ell^m(\theta)}{\partial\theta}\begin{cases}\cos m\phi \\ \sin m\phi\end{cases}\hat{\boldsymbol{\theta}} \\ + \frac{m}{r\sin\theta}j_\ell(kr)P_\ell^m(\theta)\begin{cases}-\sin m\phi \\ \cos m\phi\end{cases}\hat{\boldsymbol{\phi}}\end{pmatrix} \qquad (10.4)$$

Here, $j'_\ell(\rho) = \partial j_\ell(\rho)/\partial\rho$.

The solution of Laplace's equation for IF modes is

$$\Phi_E = B_1 r^\ell P_\ell^m(\theta)\begin{cases}\cos m\phi \\ \sin m\phi\end{cases} \quad r \leq R$$

$$= B_2 r^{-(\ell+1)} P_\ell^m\begin{cases}\cos m\phi \\ \sin m\phi\end{cases} \quad r \geq R \qquad (10.5)$$

The associated electric field within the sphere is

$$\mathbf{E}_1 = -B_1 \left(\ell r^{\ell-1} P_\ell^m(\theta) \begin{Bmatrix} \cos m\phi \\ \sin m\phi \end{Bmatrix} \hat{\mathbf{r}} + r^{\ell-1} \frac{\partial P_\ell^m(\theta)}{\partial \theta} \right.$$

$$\left. \begin{Bmatrix} \cos m\phi \\ \sin m\phi \end{Bmatrix} \hat{\theta} + \frac{m r^{\ell-1}}{\sin \theta} P_\ell^m(\theta) \begin{Bmatrix} -\sin m\phi \\ \cos m\phi \end{Bmatrix} \hat{\phi} \right) \quad (10.6)$$

The external field is

$$\mathbf{E}_2 = -B_2 \left(-(\ell+1) r^{-(\ell+2)} P_\ell^m(\theta) \begin{Bmatrix} \cos m\phi \\ \sin m\phi \end{Bmatrix} \hat{\mathbf{r}} + r^{-(\ell+2)} \frac{\partial P_\ell^m(\theta)}{\partial \theta} \begin{Bmatrix} \cos m\phi \\ \sin m\phi \end{Bmatrix} \hat{\theta} \right.$$

$$\left. + \frac{m r^{-(\ell+2)}}{\sin \theta} P_\ell^m(\theta) \begin{Bmatrix} -\sin m\phi \\ \cos m\phi \end{Bmatrix} \hat{\phi} \right) \quad (10.7)$$

Satisfying the electrical boundary conditions gives

$$B_2 = -p B_1 \frac{\ell}{\ell+1} R^{2\ell+1} \quad (10.8)$$

and

$$B_1 = -\alpha A \frac{1}{R^\ell} j_\ell(kR) \frac{\ell+1}{\ell} \frac{1}{\left(p + \frac{\ell+1}{\ell}\right)} \quad (10.9)$$

where $p = \varepsilon_W(\omega)/\varepsilon_B(\omega)$ and $\alpha = e^*/V_0 \varepsilon_\infty$ as before. The polar displacement becomes

$$\mathbf{u}_{polar} = A \left(k j_\ell'(kr) - \frac{\ell r^{\ell-1}}{R^\ell} \Gamma j_\ell(kr) \right) P_\ell^m(\theta) \begin{Bmatrix} \cos m\phi \\ \sin m\phi \end{Bmatrix} \hat{\mathbf{r}}$$

$$+ A \left(1 - \left[\frac{r}{R}\right]^\ell \Gamma \right) j_\ell(kr) \frac{\partial P_\ell^m(\theta)}{r \partial \theta} \begin{Bmatrix} \cos m\phi \\ \sin m\phi \end{Bmatrix} \hat{\theta} \quad (10.10)$$

$$+ A \left(1 - \left[\frac{r}{R}\right]^\ell \Gamma \right) j_\ell(kr) \frac{m}{r \sin \theta} P_\ell^m(\theta) \begin{Bmatrix} \cos m\phi \\ -\sin m\phi \end{Bmatrix} \hat{\phi}$$

$$\Gamma = \frac{\ell+1}{\ell s \left(p + \frac{\ell+1}{\ell}\right)} \quad (10.11)$$

10.2.2 The TO Modes

The solutions for the TO modes are

$$\mathbf{A}_1 = \nabla \times (\psi \mathbf{r})$$
$$\mathbf{A}_2 = \frac{1}{k} \nabla \times \mathbf{A}_1 \qquad (10.12)$$

The scalar quantity is given by

$$\psi = j_\ell(k_T r) P_\ell^m(\theta) \begin{cases} \cos m\phi \\ \sin m\phi \end{cases} \quad \text{or} \quad i_\ell(\eta r) P_\ell^m(\theta) \begin{cases} \cos m\phi \\ \sin m\phi \end{cases} \qquad (10.13)$$

depending on whether the relevant wave vector is real or imaginary. $i_\ell(\rho)$ is a modified spherical Bessel function. The displacement of the TO2 mode turns out to be

$$\mathbf{u}_2 = C_2 j_\ell(k_T r) \left(\frac{m}{\sin\theta} P_\ell^m(\theta) \begin{cases} \cos m\phi \\ -\sin m\phi \end{cases} \hat{\theta} + \frac{\partial P_\ell^m}{\partial \theta} \begin{cases} \cos m\phi \\ \sin m\phi \end{cases} \hat{\phi} \right) \qquad (10.14)$$

This mode can satisfy the mechanical boundary conditions without hybridization with $j_\ell(k_T R) = 0$. This is not an option for the TO1 mode; it must combine with the LO mode to satisfy the boundary condition. This is also true for the LO mode, where the TO1 mode must have the frequency of the LO mode.

The displacements of the resultant triple hybrid at the LO frequency are

$$\mathbf{u} = \left(A \left(k j_\ell'(kr) - \frac{\ell r^{\ell-1}}{R^\ell} \Gamma j_\ell(kr) \right) P_\ell^m(\theta) + C_1 \frac{\ell(\ell+1)}{r} i_\ell(\eta r) P_\ell^m(\theta) \right) \begin{cases} \cos m\phi \\ \sin m\phi \end{cases} \hat{\mathbf{r}}$$

$$+ \left(A \left(1 - \left[\frac{r}{R}\right]^\ell \Gamma \right) j_\ell(kr) \frac{\partial P_\ell^m(\theta)}{r \partial \theta} + C_1 \frac{1}{r \sin\theta} \frac{\partial (r i_\ell(\eta r))}{\partial r} \frac{\partial P_\ell^m(\theta)}{\partial \theta} \right) \begin{cases} \cos m\phi \\ \sin m\phi \end{cases} \hat{\theta}$$

$$+ \left(A \left(1 - \left[\frac{r}{R}\right]^\ell \Gamma \right) j_\ell(kr) \frac{m}{r \sin\theta} P_\ell^m(\theta) + C_1 \frac{m}{r \sin\theta} \frac{\partial (r i_\ell(\eta r))}{\partial r} P_\ell^m \right) \begin{cases} \cos m\phi \\ -\sin m\phi \end{cases} \hat{\phi}$$

$$(10.15)$$

The tangential components at the surface are dominated by the evanescent TO components, which implies that $C_1 \approx 0$. In which case, the following dispersion relation holds for $u_r = 0$:

$$k_L j_\ell'(k_L R) - \frac{\ell+1}{R_s \left(p + \frac{\ell+1}{\ell} \right)} j_\ell(kR) = 0 \qquad (10.16)$$

The significant difference between the LO and TO frequencies, which leads to the TO components of the hybrid being heavily evanescent, means that equation (10.16) for the dispersion and equation (10.10) for the polar modes are excellent approximations, corresponding to the reduced mechanical boundary condition $u_r = 0$.

Evidently a hybrid at the TO frequency would require the assistance of an LO mode of very large wave vector for the LO frequency to be that of the TO mode. This would mean that the mechanical boundary condition for TO hybrids would simplify to the requirement that only the tangential components of the displacement need vanish.

10.3 Polar Double Hybrids

The components of the particle displacements of the polar mode are

$$\mathbf{u} = A \begin{pmatrix} \left(kj_\ell''(kr) - \frac{\ell r^{\ell-1}}{R^\ell}\Gamma j_\ell(kr)\right) P_\ell^m(\theta) \begin{cases} \cos m\phi \\ \sin m\phi \end{cases} \hat{\mathbf{r}} \\ + \left(1 - \left[\frac{r}{R}\right]^\ell \Gamma\right) j_\ell(kr) \frac{\partial P_\ell^m(\theta)}{r\partial\theta} \begin{cases} \cos m\phi \\ \sin m\phi \end{cases} \hat{\boldsymbol{\theta}} \\ + \left(1 - \left[\frac{r}{R}\right]^\ell \Gamma\right) j_\ell(kr) \frac{m}{r\sin\theta} P_\ell^m(\theta) \begin{cases} \cos m\phi \\ -\sin m\phi \end{cases} \hat{\boldsymbol{\phi}} \end{pmatrix} \quad (10.17)$$

and

$$\Gamma = \frac{\ell+1}{\ell s \left(p + \frac{\ell+1}{\ell}\right)} \quad (10.18)$$

The dispersion relation is

$$kj_\ell'(kR) - \frac{\ell+1}{Rs\left(p + \frac{\ell+1}{\ell}\right)} j_\ell(kR) = 0 \quad (10.19)$$

It should be noted that the condition $\ell = 0$ is exceptional in that both the external and internal electric fields disappear, and the hybrid reduces to the LO mode alone, with no help needed from the TO modes. The only condition is on u_r: $j_0'(kR) \equiv -j_1(kR) = 0$. The fields also disappear for arbitrary ℓ when $j_0(kR) = 0$. Figure 10.1 shows the dispersion of hybrids in a free-standing GaAs quantum dot.

In this case, the allowed values of the wave vector are determined by the order of the spherical Bessel function as shown in Figure 10.1. For $\ell = 0$, the modes are

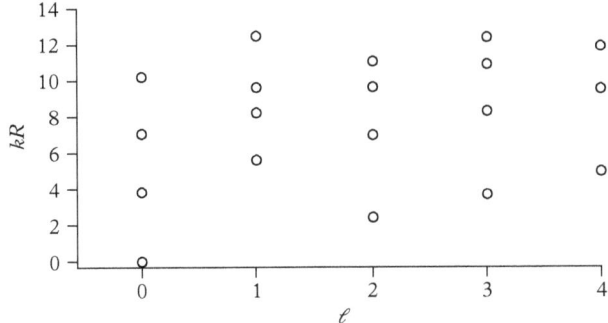

Figure 10.1 *Quantum dot hybrid modes (free-standing GaAs, R = 5 nm).*

pure LO. For $\ell > 0$, the modes are hybrids of LO and IF, but at the resonance condition the modes are IF-like. There is just one IF-like mode for each value of $\ell > 0$. In addition there are, for each ℓ, $2\ell+1$ modes characterized by the quantum number m. The total number of modes is, of course, limited by the number of unit cells in the dot.

Quantization gives

$$A^2 = \frac{1}{K^2}\left(\frac{\hbar}{2\overline{M}\omega}\right)\left[n(\omega) + \frac{1}{2} \pm \frac{1}{2}\right]$$

$$K^2 = \int \left(|u_r|^2 + |u_\theta|^2 + |u_\phi|^2\right) r^2 dr \sin\theta\, d\theta\, d\phi / (4\pi R^3/3)$$

(10.20)

Limiting attention to the case $\ell = 0$, $m = 0$, and noting that $\ell\Gamma = 0$, we obtain

$$(K^0_{0r})^2 = \frac{3q^2}{2} j_2^2(qR)\big|_{j_1(qR)=0}$$

$$(K^0_{0\theta})^2 = (K^0_{0\phi})^2 = 0$$

(10.21)

10.4 Quantum Disc and Quantum Box

The electron structure in cases of three-dimensional confinement has been studied for spheroids, ellipsoids, cylindrical discs, and square or rectangular boxes, in terms of the DC model. Here we summarize the patterns of hybrid polar modes in a quantum disc and in a quantum box.

The cylindrical disc is obtained by the addition of planar confining boundaries to the cylindrical quantum wire. The pattern of polar modes is described in terms of the scalar potential:

$$\Phi = A\mathcal{J}_m(k_L r) \begin{Bmatrix} \cos m\theta \\ \sin m\theta \end{Bmatrix} \begin{Bmatrix} \cos k_z z \\ \sin k_z z \end{Bmatrix} + BK_m(k_z r) \begin{Bmatrix} \cos m\theta \\ \sin m\theta \end{Bmatrix} \begin{Bmatrix} \cosh k_z z \\ \sinh k_z z \end{Bmatrix}$$
(10.22)

We note that the disc exhibits two different patterns of IF modes: one associated with the cylindrical boundary, the other with the planar boundaries.

A quantum box is obtained from the quantum well by inserting planar boundaries in the x and y directions. In this case the scalar potential becomes

$$\Phi = A \begin{Bmatrix} \cos k_x x \\ \sin k_x x \end{Bmatrix} \begin{Bmatrix} \cos k_y y \\ \sin k_y y \end{Bmatrix} \begin{Bmatrix} \cos k_z z \\ \sin k_z z \end{Bmatrix} + B \begin{Bmatrix} \cosh k_x x \\ \sinh k_x x \end{Bmatrix} \begin{Bmatrix} \cosh k_y y \\ \sinh k_y y \end{Bmatrix} \begin{Bmatrix} \cosh k_z z \\ \sinh k_z z \end{Bmatrix}$$
(10.23)

In both cases, B and the wave vectors are determined by the electric and mechanical boundary conditions at each surface and, as usual, A is determined by energy normalization.

Part 3

Electron–Phonon Interaction

11
The Interaction between Electrons and Polar Optical Phonons in Nanostructures: General Remarks

11.1 A Brief History

In the study of transport phenomena in semiconductor nanostructures, the interaction between electrons and phonons assumes a central significance because of its intrinsic nature. In principle, scattering due to impurities, lattice defects, and interface roughness can be eliminated. In practice, technical innovation regarding purification and crystal growth can make non-phonon scattering subordinate in many structures at room temperature, and likely to be dominant only at low temperatures, where the excitation of lattice vibrations is weak. But even there, the electron–phonon interaction cannot be ignored when the electrons become hot in high electric fields. The only other scattering mechanism that is not readily amenable to ideal crystal growth is that due to alloy fluctuations. Otherwise, in unalloyed binary semiconductors, the electron–phonon interaction is of central importance.

In nanostructures, both electrons and phonons are confined, and a description of the electron–phonon interaction must necessarily contain models of confinement. The confinement of electrons is determined by the potential barrier at each interface of the structure. The boundary conditions that have to be satisfied are the continuity of the wave function and its gradient. In many cases, the condition affecting the gradient of the Bloch function, $u(\mathbf{r})F(\mathbf{r})$, where $F(\mathbf{r})$ is an envelope function, is satisfactorily replaced by $m^{*-1}dF(z)/dz$, where m^* is the effective mass and z is along the normal to the interface (Burt 1992). Or, more simply, in cases where the barrier is high, the boundary condition reduces to $F(z) = 0$. The model of electron confinement, as just outlined, is uncontroversial.

The same cannot be said of the models of confinement of optical phonons. The boundary conditions affecting acoustic modes—continuity of particle displacement and of elastic stress—are the ones of classical physics and, again, uncontroversial. The boundary conditions for optical modes are more complex, and their elucidation is relatively recent (Foreman and Ridley 1999; Ridley 2009). Prior to this, three models were proposed. The first simply ignored optical phonon confinement entirely and calculated the electron–phonon interaction using bulk modes (Riddoch and Ridley 1983). We label this the 3DP model. Surprisingly, the result proved to be close to the result for confined phonons. The second focused on the electric fields of the polar modes and redefined the medium as a dielectric continuum (DC). Following the treatment of the polar slab by Fuchs and Kliewer (1965), the DC model takes the phonon spectrum to be confined longitudinal optical (LO) modes at the zone-centre frequency plus two interface (IF) modes, the frequency of one deriving from the well material, the other from the barrier(s), all satisfying the classical electrical boundary conditions—continuity of electric potential and of normal electric displacement. Again, surprisingly, the results proved to be close to those derived from more accurate models. However, the form of the mode patterns revealed by Raman scattering were not those of the DC model. Nor, obviously, could the DC model be applied to any non-polar system, such as Si/Ge. The third model ignored electrical boundary conditions and assumed that the relevant boundary conditions were those of acoustic modes—those also of hydrodynamics (Babiker 1986). The hydrodynamic (HD) model predicted exactly those observed by Raman scattering (Zucker et al. 1984), and it was a model that was applicable to non-polar materials. The model assumed that the elastic stress for optical modes was of the same form as for acoustic modes, which proved to be true for non-polar material but not for polar material. An analysis by Akero and Ando (1989), based on the linear-chain model, showed that the correct boundary condition must include the variation of the ionic masses, which the classical boundary conditions for acoustic modes do not. It was clear that none of the three models was correct.

From the classical study of acoustic modes in a non-polar slab, it was found that the acoustic boundary conditions could not be satisfied for a single mode with components of particle displacement along the normal to the surface. For such cases it was necessary for there to be a coherent linear combination of longitudinal acoustic (LA) and transverse acoustic (TA) modes. It became clear that the satisfaction of both mechanical and electrical boundary conditions required a coherent linear combination (LC) of LO, TO, and IF modes. But the form of the mechanical boundary condition involving stress remained unclear. Fortunately, this problem could be bypassed in many cases of interest by exploiting the large disparity of frequency and other properties between the barrier and well and assuming that the ionic displacement vanished at the interface. This allowed the correct eigenmodes to be established (Trallero-Giner et al. 1992; Nash 1992; Ridley 1992, 1993), and shown to give results in close agreement to those obtained by computer intensive microscopic modelling (Chamberlain et al. 1993). An analytical microscopic model became available (Foreman and Ridley 1999) that confirmed the dependence on mass shown by Akero and Ando and defined the mechanical boundary conditions for optical modes. The LC model with these mechanical boundary conditions, along with the usual electrical boundary conditions, is the most accurate analytical continuum model available.

11.2 Dispersion

Unlike other models, the hybrid model takes full account of lattice dispersion. Here, we take the opportunity of emphasizing the role of dispersion. In a long-wavelength approximation, it takes the frequency of the hybrid mode to be quadratic in wave vector:

$$\omega^2 = \omega_L^2 - v_L^2 q^2 \tag{11.1}$$

where ω_L is the zone-centre frequency and v_L is a velocity. In the simplest model, v_L is just the velocity of the LA mode. Dispersion has important consequences. The first is the effect on the dielectric functions associated with the reduction of frequency with wave vector. For the LO component, the permittivity becomes

$$\varepsilon(\omega) = \varepsilon_\infty \frac{\omega^2 - \omega_L^2 + v_L^2 q^2}{\omega^2 - \omega_T^2 + v_L^2 q^2} \tag{11.2}$$

where ω_T is the zone-centre TO frequency. This means that for the LO component $\varepsilon(\omega) = 0$ for all wave vectors. The fundamental dispersion for the IF component is electromagnetic, which makes, for example, in the quantum well, $q_z^2 \to -q_x^2$ in the unretarded limit. However, as a coherent component of the hybrid model, it must have the frequency of the LO component. Thus

$$\varepsilon(\omega) = \varepsilon_\infty \frac{\omega^2 - \omega_L^2}{\omega^2 - \omega_T^2} = -\varepsilon_\infty \frac{v_L^2 q^2}{\omega_L^2 - \omega_T^2 - v_L^2 q^2} \tag{11.3}$$

Moreover, the electric field associated with the IF component and its ionic displacement, **u**:

$$\mathbf{E} = -\alpha_0 s(\omega) \mathbf{u} \tag{11.4}$$

$$\alpha_0 = \frac{\omega_L^2 \overline{M}}{V_0} \left(\frac{1}{\varepsilon_\infty} - \frac{1}{\varepsilon_s} \right) \tag{11.5}$$

$$s(\omega) = \frac{\omega^2 - \omega_T^2}{\omega_L^2 - \omega_T^2} \tag{11.6}$$

Here, $\overline{M}$ is the reduced mass, V_0 is the volume of the unit cell, and $\varepsilon_\infty, \varepsilon_s$ are the high-frequency and static permittivities. $s(\omega)$ is the field factor, which, unlike the case for the LO component, reduces from unity by dispersion.

For $q = 0$, the dielectric function vanishes for both LO and IF components. As a consequence the electric field at the interface is zero when $\mathbf{u} = 0$, and there is no leakage of fields from well to barrier and vice versa. A significant consequence of dispersion is the generation of fields at the interface, even though $\mathbf{u} = 0$, because of the difference of dielectric function and field factor between the LO and IF components. This difference introduces two aspects of the hybrid model regarding the interaction with electrons—LO/IF-like and pure IF-like. The LO/IF-like interaction occurs at all frequencies, determined by dispersion, with the IF interaction opposed to that of the LO; the pure IF-like interaction occurs only at a particular frequency, determined by the lattice dispersion. In the case of the single heterostructure, for example, it is the frequency at which the sum of the permittivities of well and barrier equals zero, which occurs typically at a large wave vector.

Without dispersion, the TO component could not exist. The TO dispersion in the quadratic approximation is

$$\omega^2 = \omega_T^2 - v_T^2 q^2 \tag{11.7}$$

Coherence with the LO mode is obtained at the wave vector given by

$$\omega^2 = \omega_T^2 - v_T^2(q_{Tz}^2 + q_x^2) = \omega_L^2 - v_L^2 q^2$$

$$q_{Tz}^2 = -\frac{1}{v_T^2}[\omega_L^2 - \omega_T^2 - v_L^2 q_z^2 - (v_L^2 - v_T^2)q_x^2] = -\eta^2 \tag{11.8}$$

The z component must be imaginary and, typically, very large. The TO component is therefore evanescent and severely confined to the interface. As it has no electrical activity, and away from the surface its amplitude becomes infinitesimal, its action in the mechanical boundary condition is simply to ensure that $u_x = 0$—its mechanical energy can be ignored. In this case, the effective boundary condition reduces to $u_z = 0$. This provides a useful simplification. Nevertheless, dispersion complicates the calculation of the electron scattering rate in calling for numerical analysis. This is one reason why simpler models—the 3DP and DC—retain popularity.

11.3 Coupled Modes and Hot Phonons

However, those simple models are inadequate for treating the important practical cases where high electron densities and high electric fields introduce the phenomena of coupled plasmon–phonon modes and hot phonons. Plasmon coupling introduces extra dispersion, and high fields produce hot electrons, which, in turn, forces the phonon population to deviate from thermodynamic equilibrium. Any description of these effects will have to take into account variations depending on

Table 11.1 *Models for electrons.*

	Confined	Degenerate
	no	no
electrons		yes
	yes	no
		yes

Table 11.2 *Models for phonons.*

	Confined	Hot	Screened
		no	no
	no		yes
		yes	no
phonons			yes
		no	no
	yes		yes
		yes	no
			yes

frequency and wave vector, and this precludes simple models and calls for the hybrid approach. Also, of course, the statistics of the electron gas has to be taken into account—degenerate or non-degenerate. Although the hybrid model has general application, in practice, it has been used with the assumption that the ionic displacement vanishes at the interface. This assumption becomes questionable in systems such as the $Ga_{1-x}Al_xAs$ nanostructure when x is small. In this case, the correct mechanical boundary conditions as given by microscopic models must be used.

Tables 11.1 and 11.2 indicate the choices for models of the electron–phonon interaction, 22 in all.

Given high electric fields, the electrons will be hot and describable in terms of non-degenerate statistics, in which case the most realistic model will include confined, non-degenerate electrons interacting with confined hot and screened phonons.

12
Electrons

12.1 Confinement

In the case of electrons, it was well known what the connection rules had to be, namely, the continuity of the amplitude of the wave function and the continuity of its gradient, ensuring the continuity of electric current. However, the wave function in a crystal is a Bloch function, a combination of cell periodic wave functions and an envelope function, and very complicated. It was not easy to see how those basic connection rules could be applied. In bulk material, the electron is described very usefully in the effective-mass approximation, in which the envelope function is the solution of the Schrödinger equation:

$$\left(-\frac{\hbar^2}{2m^*}\nabla^2 + V(\mathbf{r})\right) F(\mathbf{r}) = (E - E_n) F(\mathbf{r}) \qquad (12.1)$$

where m^* is the effective mass of the electron as determined by the band structure, $F(\mathbf{r})$ is the envelope function, and E_n is the energy of the conduction band edge. In the vicinity of an interface normal to the z direction, the equation becomes

$$\left(-\frac{\hbar^2}{2}\nabla \cdot \left[\frac{1}{m^*(z)}\nabla\right] + V(\mathbf{r})\right) F(\mathbf{r}) = (E - E_n(z)) F(\mathbf{r}) \qquad (12.2)$$

Integration across an infinitesimally thin interface layer leads to connection rules, namely, the continuity of

$$F(\mathbf{r}) \text{ and } \frac{1}{m^*}\frac{dF(\mathbf{r})}{dz} \qquad (12.3)$$

which implies that if the effective mass changes then the gradient of the envelope function will be discontinuous. This connection rule works very well in practice in that it predicts the observed quantized energies of confined electrons, but it appears to contradict the fundamental quantum mechanical rule. This contradiction was resolved by Burt (1988) who applied effective mass theory to the whole of the structure, interface and all, and showed that, provided the envelope function was slowly varying in the sense that its Fourier components did not lie outside the Brillouin zone, the true Bloch function gradient was indeed continuous at the microscopic scale, but appeared discontinuous at the mesoscopic scale (Burt 1992). Thus, in most cases equation (12.2) is directly applicable. Where it may fail is when interband coupling is strong, as it is in narrow-band semiconductors (Foreman 1995).

Nanostructures consisting of at least two different semiconductors offer the possibility of limiting the motion of an electron by the presence of a potential barrier derived from the difference of electron affinity between the two materials. Limiting the motion leads to quantization of the electron's energy and leads to the description of nanostructures in terms of quantum wells, quantum wires, and quantum dots.

In bulk material, the wave function of the electron is the Bloch function:

$$\Psi(\mathbf{r}) = Au(\mathbf{r})\psi(\mathbf{r}) \qquad (12.4)$$

where $u(\mathbf{r})$ is the cell-periodic function and $\psi(\mathbf{r})$ is an envelope function, determined by the Schrödinger equation:

$$\left(-\frac{\hbar^2}{2}\nabla \cdot \frac{1}{m^*}\nabla + V(\mathbf{r})\right)\psi(\mathbf{r}) = E\psi(\mathbf{r}) \qquad (12.5)$$

where m^* is the effective mass and $V(\mathbf{r})$ is the electric potential. Confined in a nanostructure, the wave function has to satisfy boundary conditions that strictly entail the continuity of the wave function and its slope. When the effective mass varies across the boundary, the boundary condition becomes, effectively, the continuity of

$$\psi(\mathbf{r}) \text{ and } \frac{1}{m^*}\frac{\partial \psi(\mathbf{r})}{\partial z} \qquad (12.6)$$

where z is the normal to the boundary. As mentioned before, Burt (1992) has shown that the strictly true conditions prevail at the microscopic level, but that equation (12.6) describes the variation accurately at the mesoscopic level. There is no incompatibility. The effective-mass boundary conditions can be used in most cases in practice. The exception is where narrow-band semiconductors are involved, since there will be interband elements affecting the wave function. The boundary conditions set the conditions under which the electron is confined and under which the electron energy is quantized. The extent of the confinement is determined by the potential barriers in the nanostructure.

A general problem concerns the potential in the nanostructure. In the simplest cases it is a constant but, in practice, there may be depletion or accumulation at the boundaries as a result of electric fields generated by the pinning of the Fermi

126 Electrons

level at the boundaries. In the case of the single heterostructure, the field is often deliberately engineered to produce an accumulation, as it is in the HEMT, for example, or it may be spontaneously produced, as it is in the case of hexagonal crystals like GaN.

12.1.1 The Single Heterostructure

The wave function for the ground state in these cases is often modelled by the Fang–Howard function (1967), which assumes an infinite barrier:

$$\psi(x,z) = \left(\frac{b^3}{2\sigma}\right)^{1/2} e^{ik_x x} z e^{-bz/2} \qquad b = \left(\frac{33e^2 m^* N_s}{8\varepsilon_s \hbar^2}\right)^{1/3} \tag{12.7}$$

where σ is the area of the interface, eN_s is the surface density of attractive charge, and ε_s is the static permittivity. In the case of depletion, the potential can be somewhat crudely modelled by regarding it as constant beyond a certain distance from the boundaries, where the electrons are to be found.

12.1.2 The Quantum Well

Assuming infinitely high barriers and a constant potential in the well, we can write the normalized wave function as follows:

$$\psi(x,z) = \left(\frac{2}{\sigma L}\right)^{1/2} e^{ik_x x} \begin{cases} \cos k_z z & \text{symmetric} \\ \sin k_z z & \text{antisymmetric} \end{cases} \quad -L/2 \le z \le L/2$$

$$k_z = n\pi/2 \begin{cases} n \text{ odd symmetric} \\ n \text{ even antisymmetric} \end{cases} \tag{12.8}$$

12.1.3 The Quantum Wire

Once again, we assume an infinitely high barrier and a constant potential in the well. The wave function has the form

$$\psi(r,z) = A e^{ik_z z} \mathcal{J}_m(k_n r) \begin{cases} \cos m\theta \\ \sin m\theta \end{cases} \quad m = 0, \pm 1, \pm 2 \ldots \quad 0 \le z \le L, \quad 0 \le r \le R$$

$$A^2 \int_0^{R,2\pi,L} \mathcal{J}_m^2(k_n r) \begin{cases} \cos^2 m\theta \\ \sin^2 m\theta \end{cases} r\, dr\, d\theta\, dz = 1 \tag{12.9}$$

where $\mathcal{J}_m(k_n r)$ is a Bessel function of the first kind.

The integrations are

$$\int_0^R r\mathcal{J}_m^2(k_n R)\,dr = \frac{R^2}{2}\left(\mathcal{J}_m^2(k_n R) - \mathcal{J}_{m-1}(k_n R)\mathcal{J}_{m+1}(k_n R)\right)$$

$$\int_0^{2\pi} \begin{Bmatrix} \cos^2 m\theta \\ \sin^2 m\theta \end{Bmatrix} d\theta = \begin{cases} 2\pi & m = 0 \\ \pi & m \neq 0 \end{cases}$$

$$\int_0^L dz = L \tag{12.10}$$

At the boundary of the wire, the wave function must vanish: $\mathcal{J}_m(k_n R) = 0$, and k_n is determined by the zeros of the Bessel function, labelled by the number n. Among the properties of the Bessel function is the relation

$$\mathcal{J}_{m-1}(\rho) + \mathcal{J}_{m+1}(\rho) = \frac{2m}{\rho}\mathcal{J}_m(\rho) \tag{12.11}$$

Since $\mathcal{J}_m(k_n R) = 0$, $\mathcal{J}_{m-1}(k_n R) = -\mathcal{J}_{m+1}(k_n R)$, and so

$$\int_0^R r\mathcal{J}_m^2(k_n R)\,dr = \frac{R^2}{2}\mathcal{J}_{m+1}^2(k_n R) \tag{12.12}$$

The wave function is

$$\psi_{n,m}(r,z) = \left(\frac{1}{\pi R^2 L \mathcal{J}_{m+1}^2(k_n R)}\right)^{1/2} \times \begin{Bmatrix} 2\ (m \neq 0) \\ 1\ m = 0 \end{Bmatrix}^{1/2} e^{ik_z z} \mathcal{J}_m^2(k_n r) \begin{Bmatrix} \cos m\theta \\ \sin m\theta \end{Bmatrix}$$

$$\tag{12.13}$$

12.1.4 The Quantum Dot

For an infinitely high potential barrier and a constant potential, the wave function for a single electron is

$$\psi(r,\theta,\phi) = Aj_\ell(k_n r) P_\ell^m(\theta) \begin{Bmatrix} \cos m\phi \\ \sin m\phi \end{Bmatrix} \quad m = -\ell,\ -\ell+1,\ \ldots 0 \ldots \ell-1,\ \ell$$

$$\tag{12.14}$$

$$A^2 \int_{0,-\pi,0}^{R,\pi,2\pi} j_\ell^2(k_n r)[P_\ell^m(\theta)]^2 \begin{Bmatrix} \cos^2 m\phi \\ \sin^2 m\phi \end{Bmatrix} r^2\,dr\sin\theta\,d\theta\,d\phi = 1$$

Here, $j_\ell(k_n r)$ is the spherical Bessel function and $P_\ell^m(\theta)$ is the associated Legendre function. In many practical cases of interest, the electron is accompanied by a hole, and, consequently, there is a coulomb interaction. In view of this, for brevity we limit attention to the ground state of the single electron:

$$\psi_{10}(r,\theta,\phi) = A j_0(k_1 r)$$

$$A^2 4\pi \int_0^R j_0^2(k_1 r) r^2 dr = 1 \qquad (12.15)$$

Noting that $j_\ell(\rho) = \left(\frac{\pi}{2\rho}\right)^{1/2} \mathcal{J}_{\ell+1/2}(\rho)$, and that $j_0(k_1 R) = 0$, we can perform the integral in the same way as for the quantum wire, and obtain

$$\psi_{1,0}(r,\theta,\phi) = \left(\frac{1}{2\pi R^3 j_0^2(k_1 R)}\right)^{1/2} j_0(k_1 r) \qquad (12.16)$$

12.2 Scattering Rate

In what follows, the focus will be limited to the lifetime of the electron in a given state as determined by the scattering rate as given by the Fermi Golden Rule:

$$W = \frac{2\pi}{\hbar} \sum |H_{\text{int}}|^2 \delta(E_{\text{final}} - E_{\text{initial}}) \qquad (12.17)$$

$$H_{\text{int}} = \psi_{\mathbf{k}'}^* e\phi(\mathbf{q}) \psi_{\mathbf{k}} \qquad (12.18)$$

where $\phi(\mathbf{q})$ is the electrical potential associated with the polar optical hybrid. This can be derived from the lattice and electric scalar potentials.

A comment on nomenclature: the symbol for the wave vector is commonly taken to be k. When both electrons and phonons are involved, we will retain k for the electron wave vector, and use q for the phonon wave vector.

The following chapters describe the scattering rate by polar hybrid phonons in single heterostructures, quantum wells, and quantum wires. Technology is more concerned with the corresponding mobilities, since these determine the transport properties. The evaluation of mobilities entails the solution of the Boltzmann equation to give the scattering rate as a function of electron energy. An exact method of solution involves a ladder technique similar to that used by Delves (1959). This technique has been described at length by Fletcher and Butcher (1972). An account can be found in Ridley (2013). Application to the case of a GaN quantum well has been given by Anderson and colleagues (2001), assuming bulk-like LO modes. A similar evaluation involving hybrid modes has yet to be carried out.

13
Scattering Rate in a Single Heterostructure

13.1 Scattering Rate

We assume reduced boundary conditions (see Section 7.4) and elastic isotropy, and focus on the interaction with polar optical phonons. We take the heterostructure to consist of a barrier and well, unbounded in the x direction and situated in the region $L \leq z \leq L$, L large, with the interface at $z = 0$.

The scattering potential associated with hybrid LO modes in the well (Section 7.4) is

$$\phi(\mathbf{q}, \mathbf{r}) = i\alpha_0 A e^{i(q_x x - \omega t)} \left(\cos q_z z - \Gamma \sin q_z z - s p e^{-q_x z} \right)$$

$$\alpha_0 = \frac{e^*}{V_0 \varepsilon_\infty}, \quad e^{*2} = \omega_T^2 \overline{M} V_0 (\varepsilon_s - \varepsilon_\infty), \quad A = \frac{1}{Q} \left(\frac{\hbar}{2\overline{M}\omega} \right)^{1/2} (a_{\mathbf{q}} + a^\dagger_{-\mathbf{q}}) \quad (13.1)$$

$$Q^2 = \frac{1}{2}(q_x^2 + q_z^2)(1 + \Gamma^2), \quad \Gamma = p(q_x/q_z), \quad p = [s(1+r)]^{-1}$$

Here, s is the field factor and r is the ratio of well to barrier permittivities. There is also a scattering potential associated with evanescent barrier modes, which we will consider later.

We limit attention to the scattering within the lowest subband of electron energies in the well and assume the electron wave function to be of the Fang–Howard form (equation (12.1)). The rate involves sums over the phonon modes and final electron states:

$$W = \frac{2\pi}{\hbar} \frac{1}{N} \sum_{\mathbf{q}} \int |H_{\text{int}}|^2 \delta(E' - E \pm \hbar\omega) N(E') dE' \quad (13.2)$$

where N is the number of unit cells and E is the kinetic energy of the electron in the lowest subband. The interaction Hamiltonian is

$$H_{\text{int}} = \int \psi^*(\mathbf{r}) e\phi(\omega, \mathbf{q}, \mathbf{r}) \psi(\mathbf{r}) d\mathbf{r} \tag{13.3}$$

The sum over x is unity, provided crystal momentum is conserved:

$$H_{\text{int}} = \delta_{k'_x, k_x \pm q_x} \int_0^L \psi(z)^2 e\phi(\omega, q_z, z) dz \tag{13.4}$$

where q_z is the wave vector of the LO component of the hybrid that satisfies periodic boundary conditions:

$$q_z = \pm 2\pi n/L \tag{13.5}$$

in which n is an integer. Taking L to be large, we can convert the sum over q_z in equation (13.2) to an integral provided we express the squared modulus of the interaction Hamiltonian, as follows:

$$|H_{\text{int}}|^2 = \int_{-\infty}^{\infty} dq_z \frac{L}{2\pi} \iint \psi(z')^2 \phi(\omega, q_z, z') \psi(z)^2 \phi(\omega, q_z, z) dz' dz \tag{13.6}$$

Setting the limits of q_z to $\pm\infty$ should incur little error since the strength of the interaction falls off rapidly with q_z. If more accuracy is required, the limits must be the zone boundaries.

It is convenient to express the potential with energy normalisation as follows:

$$\phi(z) = i\alpha_0 \left(\frac{\hbar}{2\bar{M}\omega}\right)^{1/2} f(q_x, q_z, z)(a_q + a^{\dagger}_{-q})$$

$$f(q_x, q_z, z) = \frac{1}{Q(q_x, q_z)} \left(\cos q_z z - \Gamma \sin q_z z - spe^{-q_x z}\right) \tag{13.7}$$

This extracts the dependence on q_z and z. Converting the sum to an integral over q_z and extending the limits to $\pm\infty$ enables us to write

$$G(q_x, q_z, z', z) = \int_{-\infty}^{\infty} f(q_x, q_z, z') f(q_x, q_z, z) dq_z L/2\pi = G_1(q_x) G_2(z', z) \tag{13.8}$$

The integral over z defines a form factor

$$F(q_x) = \iint \psi(z')^2 \psi(z)^2 G_2(z', z) dz' dz \tag{13.9}$$

Taking into account the conservation of crystal momentum, the delta function is

$$\delta\left(\frac{\hbar^2 k'^2}{2m^*} - \frac{\hbar^2 k^2}{2m^*} \pm \hbar\omega_L\right) = \delta\left(\frac{\hbar^2 q_x^2}{2m^*} \pm \frac{\hbar^2 k q_x}{m^*}\cos\theta \mp \hbar\omega_L\right) \quad (13.10)$$

The upper sign is for absorption, the lower for emission. The phonon wave vector is constrained to lie between q_{min} and q_{max}:

$$q_{min} = k[(1 + \hbar\omega_L/E_k)^{1/2} - 1] \quad \text{absorption}(\theta = 0)$$
$$q_{max} = k[(1 + \hbar\omega_L/E_k)^{1/2} + 1] \quad \text{absorption}(\theta = \pi)$$
$$q_{min} = k[1 - (1 - \hbar\omega_L/E_k)^{1/2}] \quad \text{emission}(\theta = 0)$$
$$q_{max} = k[1 + (1 - \hbar\omega_L/E_k)^{1/2}] \quad \text{emission}(\theta = 0)$$

(13.11)

The integral over the angle is changed according to

$$\int_0^{2\pi} d\theta \to 2\int_{-1}^{1} d(\cos\theta)/\sin\theta \quad (13.12)$$

The scattering rate then takes the form

$$W = W_0 \frac{1}{2}\left(\frac{2m^*\omega_L}{\hbar}\right)^{1/2} \frac{\omega_L}{\omega}\left[n(\omega) + \frac{1}{2} \pm \frac{1}{2}\right] \frac{1}{k} \int G_1(q_x) F(q_x) \frac{dq_x}{\sin\theta(k, q_x)} \quad (13.13)$$

$$W_0 = \frac{e^2}{4\pi\hbar\varepsilon_p}\left(\frac{2m^*\omega_L}{\hbar}\right)^{1/2}$$

$$\frac{1}{\varepsilon_p} = \frac{1}{\varepsilon_\infty} - \frac{1}{\varepsilon_s}$$

(13.14)

Here, m^* is the electron effective mass, n is the phonon number, k is the electron wave vector, and θ is the scattering angle.

There are two aspects of the LC or hybrid model that depend on the value of r, the ratio of the permittivities. Being a function of frequency, r varies with the dispersion; for long waves $r \sim 0$ and the LC is LO-like, but at and near the critical condition $r = -1$, the LC is IF-like. Clearly, an accurate assessment of the overall rate calls for numerical work, but a useful insight can be obtained from an approximate analytical model that exploits the weakening of the interaction with electrons towards large wave vectors. For small wave vectors, we may put $r \sim 0$,

$p \sim 1$, and $s \sim 1$, essentially ignoring dispersion entirely. The LC is then LO-like and the integration over q_z gives

$$G(q_x, q_z, z', z) = \frac{L}{2q_x}\left(e^{-q_x|z'-z|} - e^{-q_x|z'+z|}\right) \qquad (13.15)$$

Thus, for the LO-like case:

$$G_1(q_x) = \frac{L}{2q_x}, \quad G_2(z', z) = e^{-q_x|z'-z|} - e^{-q_x|z'+z|} \qquad (13.16)$$

from which the form factor can be calculated. The first term in the expression for G_2 is that obtained for bulk phonons, the second for IF modes. Because the z component of the IF mode displacement is opposed to that of the LO component, the accompanying fields are also opposed, hence the minus sign. Moreover, in the absence of dispersion there is no remote-phonon effect and the IF contribution is simply to counter the effect of the LO mode (see Fig. 13.1).

For $r \sim -1$, the integration is less straightforward—$p \to \infty$, and the modes contributed by the barrier become important. The condition occurs at two frequencies, $\omega_\pm$, the upper frequency being barrier-like, the lower well-like (see Fig. 13.2).

In either case, the wave vectors at which the condition occurs are typically very large. The conservation of crystal momentum restricts q_x to be typically of order $q_0 = (2m^*\omega/\hbar)^{1/2}$, which is relatively small, in which case q_z is typically very large, say q_{z-} associated with the lower frequency, q_{z+} associated with the upper frequency. The scattering potential is then dominated by the well and barrier IF components, each having the form

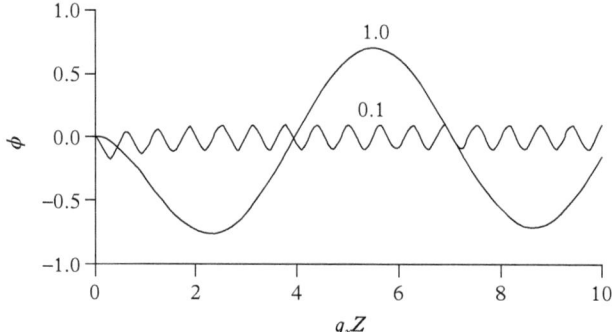

Figure 13.1 *The z component of the LO-like potential (arbitrary units) in the absence of dispersion for fixed q_x. The numbers on the curves are q_x/q_z.*

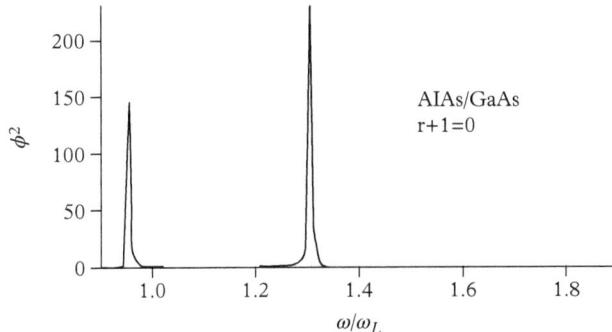

Figure 13.2 The squared potential component (s^2p^2) associated with the IF interaction for the two frequencies in the AlAs/GaAs structure for which $r + 1 = 0$.

$$f(q_x, q_z, z')f(q_x, q_z, z) = \frac{1}{Q^2(q_x, q_z)} s^2 p^2 e^{-q_x|z'+z|} \approx \frac{2s^2 e^{-q_x|z'+z|}}{(q_z^2 s^2 (1+r)^2 + q_x^2)} \quad (13.17)$$

We have assumed that $q_z \gg q_x$. We can evaluate the integral over q_z by putting $\omega^2 = \omega_\pm^2 - v_L^2 \Delta q_z^2$ and we obtain

$$1 + r = -(q_z^2 - q_{z0}^2) r_\infty v_{L\pm}^2 T_\pm^2$$

$$T_\pm^2 = \pm \left[\frac{2\omega_\pm^2 - \omega_{LB}^2 - \omega_{TW}^2}{(\omega_\pm^2 - \omega_{LB}^2)(\omega_\pm^2 - \omega_{TW}^2)} - \frac{2\omega_\pm^2 - \omega_{LW}^2 - \omega_{TB}^2}{(\omega_\pm^2 - \omega_{LW}^2)(\omega_\pm^2 - \omega_{TB}^2)} \right]$$

$$r_\infty = \varepsilon_{W\infty}/\varepsilon_{B\infty} \quad (13.18)$$

Here, $v_\pm$ are the barrier and well velocities. Ignoring the variation in the field factor with q_z, we obtain

$$G_\pm(q_x, q_z, z'z) = \frac{L}{2q_x} \frac{s_\pm}{r_{\infty\pm} v_{L\pm}^2 q_{z0\pm}^2 |T_\pm^2|} e^{-q_x|z'+z|} = \frac{L}{2q_x} \frac{s_\pm}{r_{\infty\pm}(\omega_{L\pm}^2 - \omega_\pm^2)|T_\pm^2|} e^{-q_x|z'+z|}$$

$$r_{\infty+} = r_\infty^{-1}, \quad r_{\infty-} = r_\infty \quad (13.19)$$

Thus, for the IF–like case:

$$G_1(q_x) = \frac{L}{2q_x} \frac{s_\pm}{r_{\infty\pm}(\omega_{L\pm}^2 - \omega_\pm^2)|T_\pm^2|}, \quad G_2(z', z) = e^{-q_x|z'+z|} \quad (13.20)$$

In equation (13.13) for the RP barrier rate, ε_p and the LO frequency are quantities determined by the properties of the barrier.

In summary, we note that the overall scattering rate can be seen approximately to be the sum of three processes:

$$W_{total} = W_{WLO}|_{r\approx 0} + W_{WIF}|_{1+r\approx 0} + W_{BIF}|_{1+r\approx 0} \qquad (13.21)$$

In respect of this structure, the hybrid model and the DC model (Mori and Ando 1989) are in agreement.

We note that a polar/non-polar structure would eliminate RP scattering in the polar region.

In high-power devices, conditions are far from being at thermodynamic equilibrium. Sum rules that purport to validate the use in simple models have assumed thermodynamic equilibrium for the phonons. The use of simple models to describe the electron–phonon interaction in such cases is therefore questionable. And we note that these simple models include the LO-like and IF-like analysis of the hybrid model, whose results are encapsulated in equations (13.16) and (13.30). In practical systems the electrons become extremely hot and the phonon population is driven far from thermodynamic equilibrium. There are also coupled-mode effects. These are limited to small wave vectors where Landau damping is weak, changing over to static screening where Landau damping is strong. Hot phonon populations are far from being uniform over frequency, as a consequence of variations in emission rate and phonon lifetime. Simple models are limited to the zone-centre LO frequency and the IF frequencies, but, even then, the relevant phonon numbers are far from equilibrium values, and have to be determined in (ideally) a self-consistent way.

14
Scattering Rate in a Quantum Well

14.1 Preliminary

Combining simplicity and clarity, we focus on a structure in which a polar material is sandwiched between identical polar layers that act as barriers to electrons. The sandwiched layer is the quantum well in which electrons are completely confined in the space $-L/2 \geq z \leq L/2$. The barriers are supposed to be very wide and to contain hybrid modes similar to, but not quite identical with, those in the single heterostructure. All modes in the well and barriers are assumed to obey the mechanical boundary condition $\mathbf{u} = 0$, and the usual electric ones too. Further, taking the TO components to be extremely evanescent, we can adopt the reduced boundary condition $u_z = 0$. The scattering of electrons in the well is due to the interaction between the electrons and the hybrid LO/IF modes of the well plus the IF mode contributed by the barriers, the latter introducing the remote-phonon effect.

14.2 Scattering Rate Associated with Quantum Well Modes

The rate is given by

$$W = \frac{2\pi}{\hbar} \frac{1}{N} \sum_{\mathbf{q}} \int |H_{\text{int}}|^2 \, \delta(E' - E \pm \hbar\omega) N(E') \, dE' \tag{14.1}$$

The interaction Hamiltonian is

$$H_{\text{int}} = \int \psi_f^*(\mathbf{r}) e\phi(\mathbf{q},\mathbf{r}) \psi_i(\mathbf{r}) \, d\mathbf{r} \tag{14.2}$$

where ϕ is the scattering potential. The electron wave function is given by

$$\psi(\mathbf{r}) = \frac{1}{\sigma^{1/2}} e^{i(k_x x - \omega t)} \left(\frac{2}{L}\right)^{1/2} \begin{cases} \cos k_s z \\ \sin k_a z \end{cases} \quad (14.3)$$

where σ is the area of the interface and L is the width of the well. For fully confined electrons $k_{s,a} = n\pi/2$; in the case of the symmetrical pattern, n is an odd integer; for the antisymmetrical pattern, n is even. In what follows, we limit attention to scattering within the ground state ($n = 1$).

The potential, derived from equations (8.12) and (8.16), can be written as

$$\phi(q_x, q_z, z) = i\alpha_1 e^{i(q_x x - \omega t)} f_s(q_x, q_z, z) \left(\frac{\hbar}{2M\omega}\right)^{1/2} (a_{\mathbf{q}} + a^\dagger_{-\mathbf{q}}) \quad (14.4)$$

$$f_s(q_x, q_z, z) = \frac{1}{Q_s}(\cos q_z z - s\Gamma_s \cosh q_z z)$$

where $Q_s = K_s/2^{1/2}$ and K_s is given in equation (8.13). Note that intrasubband transitions require the symmetrical hybrid. We will need to distinguish the contribution to the scattering rate from the well and the barriers, so the subscript 1 refers to the well, subscript 2 will refer to the barriers. Integrating over the area of the interface gives

$$H_{\text{int}} = \delta_{k', k \pm q_x} \int_{-L/2}^{L/2} \psi^2(z) e\phi(q_x, q_z, z) dz \quad (14.5)$$

Crystal momentum is conserved in the plane.

The scattering rate associated with a frequency ω can be expressed as follows:

$$W = W_1 \left(\frac{2m^* \omega_{L1}}{\hbar}\right)^{1/2} \frac{\omega_{L1}}{\omega} [n(\omega) + 1/2 \pm 1/2] \frac{1}{kL} \sum_{q_z} \int_{q_{x\min}}^{q_{x\max}} G_1^2(q_x, q_z) \frac{dq_x}{\sin\theta(k, q_x)}$$

$$W_1 = \frac{e^2}{4\pi\hbar\varepsilon_1} \left(\frac{2m^* \omega_{L1}}{\hbar}\right)^{1/2} \frac{1}{\varepsilon_1} = \frac{1}{\varepsilon_{\infty 1}} - \frac{1}{\varepsilon_{s1}}$$

(14.6)

The in-plane wave vector, q_x, must lie between limits determined by the conservation of momentum and energy, and generally is of the order of the wave vector of the scattered electron. At the emission threshold it is

$$q_0 = (2m^* \omega/\hbar)^{1/2} \quad (14.7)$$

The matrix element, G, often referred to as the form factor, is given by

$$G_1(q_x, q_z) = \frac{8\pi^2}{LQ}\left[\frac{\sin(q_z L/2)}{q_z[4\pi^2 - (q_z L)^2]} - \frac{s\Gamma_s \sinh(q_x L/2)}{q_x[4\pi^2 + (q_x L)]}\right] \quad (14.8)$$

Energy normalization gives

$$Q_s^2 = \frac{1}{2}\left[q_x^2 + q_z^2 + (q_x^2 - q_z^2)\frac{\sin q_z L}{q_z L} + 2\Gamma_s^2 q_x^2 \frac{\sinh q_x L}{q_x L} - 4\Gamma_s \frac{\sinh(q_x L/2)}{q_x L/2}\cos(q_z L/2)\right] \quad (14.9)$$

The dispersion relation determines the allowed values for q_z and also the character of the interaction, making it LO-like or IF-like. For the symmetrical mode equation (8.6):

$$\tan(q_z L/2) = -(q_x/q_z)p_1$$
$$p_1 = \frac{1}{s(\coth(q_x L/2) + r_1)} \quad (14.10)$$

The crucial factor is $\chi = \coth(q_x L/2) + r_1$, which is the quantum well equivalent of the single heterostructure's $1+r$. For frequencies near ω_{L1}, $r_1 \sim 0$; for narrow wells, $q_x L/2$ is typically small, and so χ is large. In this case, p_1 is small and $q_z L/2 \approx n\pi$. Such hybrid modes are LO-like.

For LO-like scattering, the form factor, noting that from equation (8.5) $\Gamma_s \sinh(q_x L/2) = p_1 \cos(q_z L/2)$, and $\frac{\sin(q_z L/2)}{4\pi^2 - (q_z L)^2} \approx \frac{1}{8\pi}$, is given by

$$G_1(q_x, q_z) = \left(\frac{L^2}{2(4\pi^2 n^2 + (q_x L)^2)}\right)^{1/2} \frac{1}{n}\left[\delta_{n,1} - \frac{16\pi^2 n s p_1 \cos n\pi}{q_x L[4\pi^2 + (q_x L)^2]}\right] \quad (14.11)$$

In the limit $q_x L \sim 0$ and $p_1 = \tanh(q_x L/2)$ and the sum over q_z, noting that $\sum_n \frac{1}{n^2} = \frac{\pi^2}{6}$, gives

$$\sum_{q_z} G_1^2(q_x, q_z) \approx \frac{L^2}{8\pi^2}\left(5 + 4\frac{\pi^2}{6}\right) \quad (14.12)$$

Substitution into equation (14.6) gives the LO-like rate (Constantinou and Ridley 1994).

The LO-like rate is given by $\chi \to \infty$; the IF-like rate is given by $\chi \approx 0$, which occurs for two frequencies; one from the barrier, one from the well:

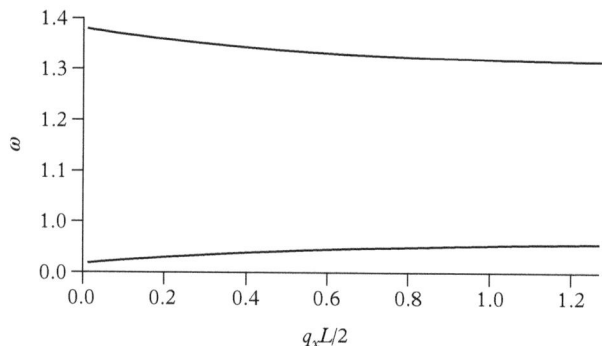

Figure 14.1 Frequencies of the IF modes in units of the LO well frequency in the case AlAs/GaAs. For small $q_xL/2$ the frequencies equal the barrier LO or well TO values (1.38 or 0.920). For large $q_xL/2$ they approach those for the single heterostructure (1.31 or 0.956).

$$\omega^2 = \frac{1}{2\{r_\infty + \coth(q_xL/2)\}}$$

$$\times \begin{pmatrix} r_\infty(\omega_{L1}^2 + \omega_{T2}^2) + \coth(q_xL/2)(\omega_{L2}^2 + \omega_{T1}^2) \\ \pm [\{r_\infty(\omega_{L1}^2 + \omega_{T2}^2) + \coth(q_xL/2)(\omega_{L2}^2 + \omega_{T1}^2)\}^2 - \\ 4\{r_\infty + \coth(q_xL/2)\}\{r_\infty\omega_{L1}^2\omega_{T2}^2 + \coth(q_xL/2)\omega_{L2}^2\omega_{T1}^2\}]^{1/2} \end{pmatrix}$$

(14.13)

(See Fig. 14.1.)

In the case that $q_xL/2$ is large, the frequencies are those of the single heterostructure. To first order in $q_xL/2$ when the latter quantity is small, the frequencies are

$$\omega_+^2 = \omega_{L2}^2 - r_{\infty 1} q_x L/2 \left[\frac{(\omega_{L2}^2 - \omega_{L1}^2)(\omega_{L2}^2 - \omega_{T2}^2)}{\omega_{L2}^2 - \omega_{T1}^2} \right]$$

$$\omega_-^2 = \omega_{T1}^2 - r_{\infty 1} q_x L/2 \left[\frac{(\omega_{T2}^2 - \omega_{T1}^2)(\omega_{L1}^2 - \omega_{T1}^2)}{\omega_{L2}^2 - \omega_{T1}^2} \right]$$

(14.14)

No problem exists for the upper frequency, but for the lower to be realizable, the dispersion must be large enough for the LO and TO bands to overlap.

If the lower frequency can be achieved, we return to the evaluation of the form factor, equation (14.8), now with $\sin(q_zL/2) = -n\pi/2$, n odd, and equation (14.9) becomes

$$Q_1^2 = 2\left(\frac{n\pi}{L}\right)\frac{\coth(q_xL/2)}{q_xL/2}$$

(14.15)

Taking $q_x L/2$ to be small,

$$G_1^2(q_x, q_z) = \frac{2L^2 s_1^2 \tanh(q_x L/2)}{(q_x L/2)^3 [4 + (q_x L/\pi)^2]} = \frac{r_{\infty 1}^2 L^2}{2} \frac{(\omega_{T2}^2 - \omega_{T1}^2)}{(\omega_{L2}^2 - \omega_{T1}^2)} \qquad (14.16)$$

where we have included the frequency dependence of the field factor, s_1. Substitution into equation (14.6) gives the IF-like rate. It may be noted that the situation where the frequency of the IF-like hybrid is close to the TO frequency entails a revision of the assumptions of a rapidly evanescent TO component of the hybrid.

14.3 Scattering Rate Associated with Barrier Modes

The situation with barrier IF modes is more straightforward, since only a modest amount of dispersion is involved. We take the field in the well at some barrier frequency to have the symmetrical pattern:

$$E_x = q_x F \cosh q_x z$$

$$E_z = -i q_x F \cosh q_x z \qquad (14.17)$$

The result from the single heterostructure, equation (7.9), is modified to produce

$$F = \alpha_2 s_2 r_2 p_2 A_2$$

$$p_2 = \frac{1}{s_2 [r_2 \coth((q_x L/2) + 1]} \quad r_2 = \frac{\varepsilon_2(\omega)}{\varepsilon_1(\omega)} \quad s_2 = \frac{\omega^2 - \omega_{T2}^2}{\omega_{L2}^2 - \omega_{T2}^2} \qquad (14.18)$$

We note that F can be written as

$$F = \alpha_2 s_1 p_1 A_2 \qquad (14.19)$$

The scattering potential is

$$\phi_2 = i\alpha_2 s_1 p_1 \frac{1}{Q_2} \left(\frac{\hbar}{2\bar{M}_2 \omega}\right)^{1/2} (a_q + a_{-q}^\dagger) \cosh q_x z \qquad (14.20)$$

The condition $\coth(q_x L/2) + r_1 = 0$ implies that $r_2 p_2 \to \infty$.
The scattering rate can be written as

$$W = W_2 \left(\frac{2m^* \omega_{L2}}{\hbar}\right)^{1/2} \frac{\omega_{L2}}{\omega} [n(\omega) + 1/2 \pm 1/2] \frac{1}{kL} \sum_{q_z} \int_{q_x \min}^{q_x \max} G_2^2(q_x, q_z) \frac{dq_x}{\sin \theta(k, q_x)}$$

$$(14.21)$$

The energy normalization has to take into account the mechanical energy in the well associated with the barrier fields. The ionic displacement is related to the field according to

$$\mathbf{u} = \frac{1}{\alpha_1 s_1 q_x} \mathbf{E} \tag{14.22}$$

The extra mechanical energy is the sum of the equal contributions from the two barriers and makes

$$Q_2^2 = \frac{1}{2}\left((q_x^2 + q_z^2)(1 + \Gamma_2^2) + \frac{q_x^2 \alpha_2^2 p_1^2}{\alpha_1^2} \frac{\sinh q_x L}{q_x L}\right) \tag{14.23}$$

The matrix element becomes

$$G_2^2(q_x q_z) = \frac{s_1^2 p_1^2}{Q_2^2} \frac{\sinh^2(q_x L/2)}{(q_x L/2)^2}\left(1 + \frac{(q_x L)^2}{4\pi^2 + q_x L}\right) \tag{14.24}$$

It is useful to note that

$$\Gamma_2^2 = \left(\frac{q_x}{q_z} p_2\right)^2 = \left(\frac{q_x}{q_z} \frac{s_1 r_1 p_1}{s_2}\right)^2 \tag{14.25}$$

and $r_1 = -\coth(q_x L/2)$. We can write

$$Q_2^2 = \frac{1}{2} q_x^2 p_1^2 \left[\left(\frac{s_1}{s_2}\right)^2 \coth^2(q_x L/2) + \left(\frac{\alpha_2}{\alpha_1}\right)^2 \frac{\sinh q_x L}{q_x L}\right] \xrightarrow{q_x L/2 \to 0} \frac{2}{L^2}\left(\frac{s_1}{s_2}\right)^2 p_1^2 \tag{14.26}$$

whence, in the same limit,

$$G_2^2(q_x, q_z) = \frac{1}{2} s_2^2 L^2 \approx \frac{L^2}{2} \tag{14.27}$$

14.4 General Remarks

Approximations relating to the factor $q_x L/2$ have been made to produce simple expressions, but accuracy requires numerical solutions of the integral over q_x in all cases. Where the approximation is valid, the form factor, G, becomes independent of q_x, in which case the integral for threshold emission reduces to $2k$. The simplified rates can be written as follows:

$$W = W_1 \left(\frac{\hbar\omega_{L1}}{E_1}\right)^{1/2} \frac{\omega_{L1}}{\omega_-}[n(\omega) + 1/2 \pm 1/2] \begin{cases} \frac{1}{4\pi}\left(5 + 4\frac{\pi^2}{6}\right) \text{ LO} \\ \pi r_{\infty 1}^2 \frac{\omega_{T2}^2-\omega_{T1}^2}{\omega_{L2}^2-\omega_{T1}^2} \text{ IF} \end{cases}$$

$$= W_2 \left(\frac{\hbar\omega_{L2}}{E_1}\right)^{1/2} \frac{\omega_{L2}}{\omega_+}[n(\omega) + 1/2 \pm 1/2]\pi \quad IF$$

(14.28)

Here, $E_1 = \hbar^2\pi^2/2m^*L^2$ is the energy of the lowest subband. As in the case of the single heterostructure, the overall interaction is the sum of three processes, as, indeed, it is in the DC model. Non-polar barriers would eliminate the significant barrier contribution.

15
Scattering Rate in Quantum Wires

15.1 General Remarks

Electron–phonon scattering rates in quantum wires with bulk acoustic and optical modes have been studied by a number of authors. The interaction with optical modes in wires with rectangular cross section, the most commonly assumed confinement geometry, has been studied by Leburton (1984). An account of the study of the interaction within the DC model can be found in Stroscio and Datta (2001). A comprehensive analysis of electron transport in a quantum wire has been carried out by Zakhleniuk and colleagues (1996). The author is not aware of any similar studies that take into account the correct normal vibrational modes of the nanostructure. Here, we limit our description to an expression for the scattering rate with hybrid optical modes.

15.2 Scattering Rate

The scattering potential for polar optical mode hybrids is

$$\Phi(n, m, q_z) = \alpha_0 A_m^2 e^{iq_z z} \left(\mathcal{J}_m(q_n r) \begin{Bmatrix} \sin m\theta \\ \cos m\theta \end{Bmatrix} + s\Gamma_m I_m(q_z r) \begin{Bmatrix} \sin m\theta \\ \cos m\theta \end{Bmatrix} \right) \quad (15.1)$$

$$\alpha_0 = \frac{e^*}{V_0 \varepsilon_\infty}, \quad e^{*2} = \omega_T^2 \bar{M} V_0 (\varepsilon_s - \varepsilon_\infty) \quad s = \frac{\omega^2 - \omega_T^2}{\omega_L^2 - \omega_T^2}$$

$$\Gamma = \frac{\mathcal{J}_m(q_n R) K'_m(q_z R)}{I'_m(q_z R) K_m(q_z R)} \left[s \left(p - \frac{I_m(q_z R) K'_m(q_z R)}{I'_m(q_z R) K_m(q_z R)} \right) \right]^{-1}$$

$$p = \frac{\varepsilon_W(\omega)}{\varepsilon_B(\omega)}, \quad \varepsilon_{W,B}(\omega) = \frac{\omega^2 - \omega_{L W,B}^2}{\omega^2 - \omega_{T W,B}^2} \quad (15.2)$$

Here, V_0 is the volume of the unit cell, $\varepsilon_\infty, \varepsilon_s$ are respectively the high-frequency and static permittivities, and s is the field factor. The frequency ω is that of

Hybrid Phonons in Nanostructures. First Edition. B.K. Ridley. © B.K. Ridley 2017.
Published in 2017 by Oxford University Press. DOI: 10.1093/acprof:oso/9780198788362.001.0001

the hybrid and is determined by the lattice dispersion. The energy normalized amplitude is given by

$$A_m^2 = \frac{2}{Q_m^2}\left(\frac{\hbar}{2\bar{M}\omega}\right)\left[n(\omega) + \frac{1}{2} \pm \frac{1}{2}\right] \quad (15.3)$$

Here, $\bar{M}$ is the reduced mass. Q_m is a normalizing factor, which can be written in the form

$$Q_m^2 = Q_{mr}^2 + Q_{m\theta}^2 + Q_{mz}^2 \quad (15.4)$$

We specialize and consider interactions within the electron ground state. Characterized by $m = 0$,

$$Q_{0r}^2 = \left(q_L^2(\mathcal{J}_1^2 - \mathcal{J}_0\mathcal{J}_2) - q_z^2\Gamma_m^2(I_1^2 - I_0I_2) - \frac{4q_Lq_z}{(q_L^2 + q_z^2)R}\Gamma_m(q_z\mathcal{J}_1I_0 + q_L\mathcal{J}_0I_1)\right)$$

$$Q_{0\theta}^2 = 0$$

$$Q_{0z}^2 = q_z^2\left((\mathcal{J}_0^2 + \mathcal{J}_1^2) + \Gamma_m^2(I_0^2 - I_1^2) + \frac{4}{(q_L^2 + q_z^2)R}(q_z\mathcal{J}_0I_1 - q_L\mathcal{J}_1I_0)\right) \quad (15.5)$$

The argument of the Bessel functions is $q_n R$, and that of the modified Bessel functions is $k_z R$. The imposed boundary condition $\mathbf{u} = 0$ results in the dispersion relation

$$[q_n\mathcal{J}_m{'}(q_n R)) + \Gamma_m q_z I_m{'}(q_z R)]\begin{cases} s \\ c \end{cases} = 0 \quad (15.6)$$

which determines q_n. Thus, for $k_z = 0$, q_n are the zeros of $J_m'(q_n R)$. Note that in an obvious notation, $s = \sin m\theta$, $c = \cos m\theta$.

The scattering rate in terms of Fermi's Golden Rule is

$$W = \frac{2\pi}{\hbar}\int |C|^2 G^2 \delta(E_{m'} - E_m \pm \hbar\omega)\left(n(\omega) + \frac{1}{2} \pm \frac{1}{2}\right) dN(E_{m'}) \quad (15.7)$$

$$C^2 = \frac{e^2\hbar\omega_L}{V_0}\left(\frac{1}{\varepsilon_\infty} - \frac{1}{\varepsilon_s}\right)\frac{\omega_L}{\omega}$$

The form factor is

$$G = \int_0^{R,2\pi,L}\left[\mathcal{J}_m(q_n r)\begin{cases} s \\ c \end{cases} + s\Gamma_m I_m(k_z r)\begin{cases} s \\ c \end{cases}\right]\psi_f\psi_i r\,dr\,d\theta\,dz \quad (15.8)$$

For simplicity, we assume that the electrons are confined by infinitely high barriers. The electron wave function is

$$\psi(n, m, k_z) = [\pi R^2 L \mathcal{J}_{m+1}^2(k_n R)]^{-1/2} e^{ik_z z} \mathcal{J}_m(k_n r) \begin{Bmatrix} s \\ c \end{Bmatrix} \quad (15.9)$$

Here, L is the length of the cylinder. Crystal momentum in the z direction is conserved, and

$$G = \delta_{k_z'' - k_z', q_z} \frac{2}{R^2 \mathcal{J}_{m''+1} \mathcal{J}_{m'+1}} \int_0^R \left[\mathcal{J}_m(q_n r) \begin{Bmatrix} s \\ c \end{Bmatrix} + s \Gamma_m I_m(k_z r) \begin{Bmatrix} s \\ c \end{Bmatrix} \right] \mathcal{J}_{m''}(k_{n''} r)$$

$$\begin{Bmatrix} s'' \\ c'' \end{Bmatrix} \mathcal{J}_{m'}(k_{n'}) \begin{Bmatrix} s' \\ c' \end{Bmatrix} r dr \quad (15.10)$$

For scattering within the ground state, $m = m' = m''$:

$$G = \delta_{k_z'' - k_z', q_z} \frac{2}{R^2 \mathcal{J}_1^2} \int_0^R [\mathcal{J}_0(q_n r) + s \Gamma_0 I_0(k_z r)] \mathcal{J}_0^2(k_1 r) r dr \quad (15.11)$$

The rate must be determined numerically.

As in the case of the single heterostructure and the quantum well, lattice dispersion will distinguish a scattering rate that is LO-like, when Γ is small, and another that is IF-like, when $\Gamma \to \infty$. The latter case is approached when

$$p + F_m(q_z R) = 0$$

$$F_m(q_z R) = -\frac{I_m(q_z R) K_m'(q_z R)}{I_m'(q_z R) K_m(q_z R)} \quad (15.12)$$

$F_m(q_z R)$ is depicted in Figure 9.2 as a function of q_z. See also the relevant discussion in Chapter 9.

16
The Electron–Phonon Interaction in a Quantum Dot

16.1 Preamble

Arrays of quantum dots have the potential for high-power lasers and for the generation of ultra-short pulses. Theoretical concerns include the description of the electron–hole interaction in a quantum dot, how electrons and holes recombine radiatively and non-radiatively, and how excited electrons and holes relax their energies. In real arrays dots vary in size and in different arrays the dots have different shapes, which any comparison of theory with experiment must take into account. Unlike the poor experimentalist, the theoretician is not as constrained by reality, and can therefore choose models that are amenable to analysis. In this chapter we focus on the electron–phonon interaction in the isotropic material that forms a spherical quantum dot. The choice of the sphere is irresistible because of its analogy with an atom—the quantum dot seen as an atom with dimensions measured in nanometres rather than in angstrom units. This makes the quantum dot intrinsically different from other nanostructures in that, at its smallest, it reduces to an individual atom or molecule with its characteristic energy level structure, and at its largest, the energy level structure becomes the band structure of the semiconductor. Establishing the band- or energy-level structure for intermediate sizes is an ongoing and extremely challenging task, an account of which would need a book in its own right. The theory of photoluminescence in quantum dots has become distinctly esoteric and certainly beyond what passes for the expertise of the present author. Moreover, it is far from being complete. Nevertheless, it is worth exhibiting the bare bones of the problem as revealed in simple models.

16.2 Electron–Lattice Coupling

The one-electron time-independent Schrödinger equation including the electron–lattice coupling is

$$(H_0 + H_{ep})\psi_\mathbf{k}(\mathbf{r}) = E_\mathbf{k}\psi_\mathbf{k}(\mathbf{r}) \qquad (16.1)$$

We can take the interaction energy in the simplest model to be proportional to the ionic displacement (as we have done in previous chapters in connection with the Fröhlich interaction):

$$H_{ep} = \mathbf{V} \cdot \mathbf{u} \qquad (16.2)$$

where $\mathbf{V}$ is the energy per unit displacement and $\mathbf{u}$ is the relevant ionic displacement associated with a particular mode. As usual, we regard the interaction as a perturbation, which, to first order, leaves the wave function unchanged but alters the electron energy:

$$E_1 = E_0 + H_1$$
$$H_1 = \langle k | \mathbf{V} \cdot \mathbf{u} | k \rangle \qquad (16.3)$$

In the absence of the electron–lattice interaction, the lattice Hamiltonian is

$$H_{lat} = \frac{1}{N} \sum_q \left(\frac{\hat{p}_q^2}{2M} + \frac{1}{2} M \omega_q^2 \hat{u}_q \right) \qquad (16.4)$$

where $\hat{p}, \hat{u}$ are, respectively, momentum and position operators, and N is a normalizing factor equal to the number of allowed modes which is equal to the number of unit cells. With the electron–lattice interaction, we have

$$H_{lat} = \left[\frac{1}{N} \sum_q \left(\frac{\hat{p}_q^2}{2M} + \frac{1}{2} M \omega_q^2 \hat{u}_q \right) + E_1 |n\rangle \right] = E_{tot} |n\rangle \qquad (16.5)$$

The perturbed electron energy can be written in terms of the position operator:

$$E_1 = E_0 + \frac{1}{N} \sum_q V_{1q} \hat{u}_q \qquad (16.6)$$

After a small manipulation, the equation becomes

$$H_{lat} = \frac{1}{N} \sum_q \left(\frac{\hat{p}_q^2}{2M} + \frac{1}{2} \hbar \omega_q (\hat{Q}_q^2 - Q_{1q}^2) \right) |n\rangle = \left(E_{tot} - E_0 + \frac{1}{N} \sum_q \frac{1}{2} \frac{V_{1q}^2}{\hbar \omega_q (M \omega_q / \hbar)} \right) \qquad (16.7)$$

The result of the electron–lattice interaction is to displace the lattice vibrations by an amount

$$Q_{1q} = -\frac{V_{1q}}{\hbar \omega_q (M \omega_q / \hbar)^{1/2}} \qquad (16.8)$$

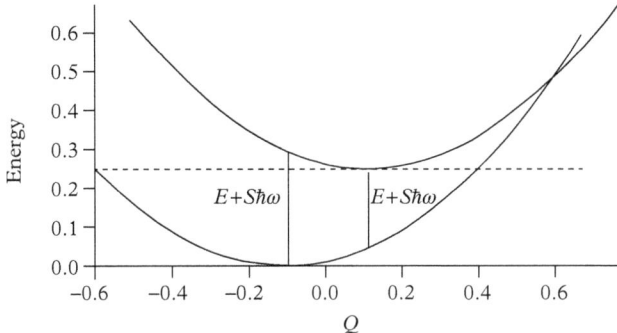

Figure 16.1 *The configuration coordinate diagram. S is the Huang–Rhys factor.*

The subscript 1 can now refer to the electron in level 1. An electron in level 2 produces a different displacement:

$$Q_{2q} = -\frac{V_{2q}}{\hbar\omega_q (M\omega_q/\hbar)^{1/2}} \qquad (16.9)$$

The effect on a transition of an electron from level 1 to level 2, or vice versa, is conveniently depicted in terms of the variation of vibrational energy with a fictional configuration coordinate Q (Fig. 16.1).

A transition between the states as a result of absorbing a photon is depicted by a vertical line, on the grounds that the heavy ions cannot respond quickly. This is an important assumption called the adiabatic approximation. Because of that, extra vibrational energy is created, which is ultimately dissipated. Equally, a downward transition leaves the system with vibrational energy to be emitted. In each case, the vibrational energy is, assuming a common frequency,

$$E_{vib} = S\hbar\omega \qquad (16.10)$$

Here, S is the factor introduced by Huang and Rhys (1950) depicting the number of phonons involved:

$$S = \frac{1}{2N} \sum_q (Q_{2q} - Q_{1q})^2 \qquad (16.11)$$

The involvement of phonons explains the Stoke's shift, the difference between the photon energy required for absorption (larger than the gap) and the photon energy emitted (less than the gap).

Because the energies of the two states are degenerate at $Q = Q_x$, it is possible for non-radiative transitions to occur. A transition would require the electron to absorb a sufficient number of phonons to reach the cross-over. In fact, second-order perturbation removes the degeneracy, but still allows the possibility of a non-radiative transition.

For an electron not in a time-independent state, the electron–phonon interaction proceeds in the usual way, with scattering rates determined by time-dependent perturbation theory. Although no current flow is involved, it is important to determine the energy relaxation rate of excited electrons.

In a quantum dot, electrons exist (1) accompanied by a hole as a result of photon absorption that induces a transition across the forbidden gap; (2) as a result of doping, deliberate or not; (3) as a result of injection, either from the barrier matrix enclosing the dot, or by tunnelling from an adjacent dot. The electron may be (4) free within the bounds of the dot; (5) bound by the coulomb attraction to the hole; (6) captured by an impurity or lattice defect. Given all these possibilities, and add to them the distribution of dot size in real systems, it is clear that the interpretation of experiments is somewhat challenging.

In what follows, we focus on the problem of the single electron-hole pair in a dot large enough for there to be a semiconductor band structure.

16.3 The Exciton

In bulk material, the energy-level structure of the electron captured by the hole is hydrogen-like. The wave function of the exciton can be taken to be the product of a travelling wave and a hydrogenic envelope function:

$$\Psi(\mathbf{r_x}, \mathbf{r}) = V^{-1/2} F(\mathbf{r}) e^{i\mathbf{K} \cdot \mathbf{r_x}}$$
$$F(\mathbf{r}) = A R_{n\ell} Y_\ell^m(\theta, \phi) \tag{16.12}$$

where A is a normalizing constant, $R_{n\ell}(\mathbf{r})$ is a radial function, and $Y_\ell^m(\theta, \phi)$ is a spherical harmonic:

$$R_{n\ell}(r) = A e^{-\rho/2} \rho^\ell L_{n+\ell}^{2\ell+1}(\rho)$$
$$Y_\ell^m(\theta, \phi) = P_\ell^m(\cos\theta) \begin{cases} \cos m\phi \\ \sin m\phi \end{cases} \tag{16.13}$$

in which $L_{n+\ell}^{2n+1}(\rho)$ is an associated Laguerre polynomial and $P_\ell^m(\cos\theta)$ is an associated Legendre polynomial. The argument of the radial function is

$$\rho = 2r/na_x, \quad a_x = \hbar^2/\mu e^2$$

Here, a_x is the excitonic radius, and μ is the reduced mass. The normalized energy eigenfunctions are

$$\psi(r,\theta,\phi) = R_{n\ell}(r) Y_\ell^m(\theta,\phi)$$

$$R_{n\ell}(r) = \left[\left(\frac{2}{na_x}\right)^3 \frac{(n-\ell-1)!}{2n[(n+\ell)!]^3}\right]^{1/2} e^{-\rho/2} \rho^\ell L_{n+\ell}^{2\ell+1}(\rho)$$

$$Y_\ell^m(\theta,\phi) = \varepsilon \left[\frac{2\ell+1}{4\pi} \frac{(\ell-|m|)!}{(\ell+|m|)!}\right]^{1/2} P_\ell^m(\cos\theta) \begin{cases} \cos m\phi \\ \sin m\phi \end{cases} \tag{16.14}$$

$$\varepsilon = \begin{cases} (-1)^m & m > 0 \\ 1 & m < 0 \end{cases}$$

Limiting attention to the 1s state, we have

$$\Psi(\mathbf{r_x},\mathbf{r}) = (\pi a_x^3)^{-1/2} e^{-r/a_x} V^{1/2} e^{i\mathbf{K}\cdot\mathbf{r_x}} \tag{16.15}$$

What is the Huang–Rhys (HR) factor associated with the exciton in the dot? A simple model for dots such that the radius of the dot (R) is larger than the exciton radius, localizes the hole at the centre of the dot, and takes the electron to be in a 1s hydrogenic orbit. The strength of the interaction with the electron (equation (16.3)) can be written as

$$V_{2q} = C(q) \int_0^\infty (\pi a_x)^{-1} e^{-2r/a_x} e^{-i\mathbf{q}\cdot\mathbf{r}} r^2 dr \sin\theta d\theta d\phi = C(q) \frac{1}{(1+(qa_x/2)^2)^2} \tag{16.16}$$

where $C(q)$ is the coupling constant. The corresponding dimensionless coordinate is

$$Q_{2q} = C(q) \frac{1}{\hbar\omega_q (M\omega_q/\hbar)^{1/2}} \frac{1}{[1+(qa_x/2)^2]^2} \tag{16.17}$$

In the initial state, the exciton does not exist, so $V_{1q} = 0$. The HR factor is

$$S = \frac{1}{2N} \sum_q C^2(q) \frac{1}{[1+(qa_x/2)^2]^4} \tag{16.18}$$

The sum can be carried out most simply by assuming that the Brillouin zone is spherical with radius q_B. The total number of modes is

$$N = \int_0^{q_B} 4\pi q^2 dq (V/(2\pi)^3) = \frac{V}{6\pi^2} q_B^3 \tag{16.19}$$

In the case of diamond and zinc blende lattices with lattice constant a_0, the elementary volume is a_0^3 containing four unit cells; in which case the number of unit cells in the volume V is $4V/a_0^3$, in which case

$$q_B = (3\pi^2)^{1/3} a_0^{-1} \tag{16.20}$$

The HR factor is then

$$S = \frac{1}{2N} \int_0^{q_B} C^2(q) \frac{1}{(\hbar\omega_q)^2 M\omega_q/\hbar} \frac{1}{[1+(qa_x/2)^2]^4} \frac{V}{2\pi^2} q^2 \, dq \tag{16.21}$$

The two major interactions are the polar and the deformation–potential interaction. In the deformation-potential case, the effect is strongest only for optical modes acting on an L valley or the valence band. For zinc blende crystals this means that only the hole would be affected. In both cases, neglecting dispersion, we can take the frequency to be the same for all modes and the reduced mass. Thus, with $x = qa_x/2$,

$$S = \frac{4}{\pi^2 \hbar\omega \overline{M}\omega^2} \left(\frac{a_0}{a_x}\right)^3 \int_0^x C^2(q) \frac{x^2}{(1+x^2)^4} \, dx \tag{16.22}$$

For the polar interaction,

$$C^2(q) = \frac{e^2}{q^2} \frac{\overline{M}\omega^2}{V_0} \left(\frac{1}{\varepsilon_\infty} - \frac{1}{\varepsilon_s}\right) \tag{16.23}$$

where V_0 is the volume of the unit cell. For the deformation interaction,[1][†]

$$C^2(q) = \left(\frac{\overline{M}}{M}\right) D^2 \tag{16.24}$$

The HR factors for the ground state are

$$S_{polar} = \frac{1}{2\pi^2} \frac{e^2}{a_x \hbar\omega} \left(\frac{1}{\varepsilon_\infty} - \frac{1}{\varepsilon_s}\right) \int_0^{q_B a_x/2} \frac{1}{(1+x^2)^4} \, dx$$

$$S_{defor} = \frac{1}{2\pi^2} \frac{D^2}{(\hbar\omega)^2 (M\omega/\hbar)} \left(\frac{a_0}{a_x}\right)^3 \int_0^{q_B a_x/2} \frac{x^2}{(1+x^2)^4} \, dx \tag{16.25}$$

---

[1] E.g. Ridley (2013), Section 3.4.

The integrals are standard. Taking the upper limit to be very much greater than unity gives for the polar case $5\pi/64$, and for the deformation case $\pi/32$.

The expression for the polar interaction refers solely to the electron, but there is also the hole. A simple model assumes that the hole is localized at $r = 0$ with charge density $+e\delta(0)$. The exciton is then similar to the shallow donor state, so it is referred to as the donor exciton. The polar interaction must reflect the difference of charge; the deformation interaction must add the contribution of the hole. We obtain

$$S_{polar} = \frac{1}{2\pi^2} \frac{e^2}{a_x \hbar \omega} \left(\frac{1}{\varepsilon_\infty} - \frac{1}{\varepsilon_s}\right) \int_0^{q_B a_x/2} \left(1 - \frac{1}{(1+x^2)^4}\right) dx \tag{16.26}$$

$$S_{defor} = \frac{1}{2\pi^2} \frac{D^2}{(\hbar\omega)^2 (M\omega/\hbar)} \left(\frac{a_0}{a_x}\right)^3 \int_0^{q_B a_x/2} \frac{x^2}{(1+x^2)^4} dx$$

The polar result is that given by Merlin and colleagues (1978). It is also the result obtained by Klein and colleagues (1990), who replace the travelling LO wave with its confined pattern in the dot. In fact, the only interaction generally considered has been the polar one. In neither case is there any of the observed dependence on dot size, which is not surprising considering the bulk-like feature of the model. But adding the confining effect of the dot makes it a three-body problem, and that means no exact solution. One begins to see difficulties.

If the coulomb interaction between the electron and hole is ignored, the polar interaction with what is now an electrically neutral system vanishes. So also does the deformation interaction with the hole, at least as regards the ground state, since the matrix element is

$$H_{ep} \sim \int_0^R j_0^2(kr) j_1(qr) r^2 dr \int_{-1}^1 \cos\theta \, d(\cos\theta) = 0 \tag{16.27}$$

Such a simple model is entirely unrealistic. The coulomb interaction cannot be ignored. Moreover, it is modified by image forces associated with the barrier material. Taking this into account, Marini and colleagues (1994) used a variational calculation to describe the donor exciton, and analysed the interaction with LO and IF modes to get the R dependence of the Huang–Rhys factor. Fedorov and Baranov (1996) used a pseudopotential to describe the exciton and calculated the interaction with confined LO modes to obtain the R dependence of the HR factor. All of the above attempts have assumed the adiabatic approximation. Non-adiabaticity has been treated by Fomin and colleagues (1998) and Klimin and colleagues (2008) using the DC theory for the electron–phonon interaction.

Spin-orbit splitting has to be included in the description of the hole state. Nor is the barrier infinitely high, and there may be more than one pair of particles. Finally, it is necessary to include the correct phonon spectrum. These considerations take this topic far beyond the remit of this book.

17
Coupled Modes

17.1 Introduction

Many practical systems, particularly high-power devices, have a high density of mobile electrons that may couple with the polar optical phonons of the system through the associated electric fields. The interaction leads to the coherent coupling of plasma and polar modes with subsequent screening or anti-screening of the polar fields. This phenomenon in bulk material has been treated by a number of authors and reviewed by Richter (1982). The treatment of coupled modes in nanostructures has tended to avoid aspects associated with confinement, essentially taking the nanostructure as bulk material (e.g. Dyson and Ridley 2012). However, in nanostructures, polar optical modes are confined and hybridized with an electric interface mode plus one or more TO modes and, furthermore, electrons are also confined. Coupled plasma-lattice modes in a nanostructure have to be described with the effects of this confinement in mind, but this must be considered work in progress.

We begin by describing coupled modes in bulk material, first in the long-wavelength approximation, which allows of a simple classical picture. The description is then extended to shorter wavelengths. Finally, the problem of coupling with hybrid modes is discussed.

17.2 Long-Wavelength Modes

Long-wavelength coupled modes in bulk material has been discussed by Kim and colleagues (1978) using diagrammatic and Green's function techniques. For our purposes here it will be sufficient to provide a simple, classical account based on the equations of motion, which, nevertheless reproduce the results of Kim and colleagues exactly (Ridley 2009).

The equation of motion that describes plasma oscillations in the absence of pressure forces and damping is

$$m^*\ddot{u}_e = -e\mathbf{E} \qquad (17.1)$$

where m^* is the effective mass of the electron, $\mathbf{u}_e$ is the displacement, and $\mathbf{E}$ is the electric field. Thus, for a wave of frequency ω, the displacement is related to the field according to

$$\mathbf{u}_e = \frac{e}{m^*\omega^2}\mathbf{E} \qquad (17.2)$$

The equation of motion for the LO mode with zero lattice dispersion and zero damping is

$$\bar{M}\ddot{\mathbf{u}}_L = -\bar{M}\omega_T^2 \mathbf{u}_L + e^*\mathbf{E} \qquad (17.3)$$

where $\bar{M}$ is the reduced mass, $\mathbf{u}_L$ is the optical displacement, ω_T is the zone-centre frequency of the TO mode, and e^* is the ionic charge (equation (3.14)):

$$e^{*2} = \bar{M}V_0\varepsilon_\infty(\omega_L^2 - \omega_T^2) \qquad (17.4)$$

where V_0 is the volume of the unit cell, ε_∞ is the high-frequency permittivity, and ω_L, ω_T are the zone-centre frequencies of the LO and TO modes. The optical displacement is related to the field according to

$$\mathbf{u}_L = -\frac{e^*}{\bar{M}(\omega^2 - \omega_T^2)}\mathbf{E} \qquad (17.5)$$

An electron wave travelling coherently along with a polar optical wave has an energy density (purely mechanical) given by

$$U = \bar{M}\omega^2 u_L^2 + nV_0 m^*\omega^2 u_e^2 \qquad (17.6)$$

where n is the electron density. Equations (17.2) and (17.5) can be used to express the electron displacement in terms of the optical displacement:

$$\mathbf{u}_e = -\frac{e\bar{M}(\omega^2 - \omega_T^2)}{e^* m^* \omega^2}\mathbf{u}_L \qquad (17.7)$$

We obtain a measure of the phonon content of the coupled mode,

$$u_L^2 = Su^2 \qquad (17.8)$$

where $\mathbf{u}$ is the effective displacement, and

$$S = \left(1 + \frac{\omega_p^2(\omega^2 - \omega_T^2)^2}{\omega^4(\omega_L^2 - \omega_T^2)}\right)^{-1} \qquad (17.9)$$

in which ω_p is the plasma frequency:

$$\omega_p^2 = \frac{ne^2}{\varepsilon_\infty m^*} \tag{17.10}$$

In order to satisfy Gauss's equation, the total dielectric function must vanish. The dielectric function associated with the electron polarization is obtained from

$$\varepsilon_e(\omega)\mathbf{E} = -ne\mathbf{u}_e \tag{17.11}$$

and equation (17.2):

$$\varepsilon_e(\omega) = -\varepsilon_\infty \frac{\omega_p^2}{\omega^2} \tag{17.12}$$

The dielectric function associated with the LO mode is obtained from

$$[\varepsilon_{LO}(\omega) - \varepsilon_\infty]\mathbf{E} = \frac{e^*}{V_0}\mathbf{u}_L \tag{17.13}$$

and equation (11.5):

$$\varepsilon_{LO} = \varepsilon_\infty \frac{\omega^2 - \omega_T^2}{\omega^2 - \omega_T^2} \tag{17.14}$$

The total dielectric function including the contribution from the core electrons is

$$\varepsilon(\omega) = \varepsilon_e(\omega) + \varepsilon_{LO}(\omega) \tag{17.15}$$

Since both plasma and LO modes are longitudinally polarized, Gauss's equation is satisfied only when the dielectric function is zero, which occurs when

$$\varepsilon_\infty \left(-\frac{\omega_p^2}{\omega^2} + \frac{\omega^2 - \omega_T^2}{\omega^2 - \omega_T^2} \right) = 0 \tag{17.16}$$

This defines the coupled-mode frequency given by

$$\omega^4 - \omega^2(\omega_L^2 + \omega_p^2) + \omega_p^2\omega_T^2 = 0 \tag{17.17}$$

$$\omega^2 = \frac{1}{2}\left(\omega_L^2 + \omega_p^2 \pm \sqrt{(\omega_L^2 + \omega_p^2)^2 - 4\omega_p^2\omega_T^2}\right) \tag{17.18}$$

The two solutions can be labelled phonon-like (ω_+) and plasmon-like (ω_-). The phonon content (equation (17.9)) can be expressed in terms of the two solutions of this equation:

$$S_+ = \frac{\omega_+^2 - \omega_p^2}{\omega_+^2 - \omega_-^2}$$

$$S_- = \frac{\omega_p^2 - \omega_-^2}{\omega_+^2 - \omega_-^2} \quad (17.19)$$

which are exactly the solutions derived by Kim and colleagues (1978) using diagrammatic techniques.

We can also define an effective charge e^{**} as follows, using

$$\mathbf{E} = \frac{e^*}{V_0(\varepsilon_{LO} - \varepsilon_\infty)} \mathbf{u}_L = \frac{e^{**}}{V_0 \varepsilon_\infty} \mathbf{u}_L$$

$$e^{**} = -e^* S_{LO} \quad S_{LO} = \frac{\omega^2 - \omega_T^2}{\omega_L^2 - \omega_T^2} \quad (17.20)$$

whence

$$(e^{**})^2 = (e^*)^2 \left(\frac{\omega^2 - \omega_T^2}{\omega_L^2 - \omega_T^2}\right)^2 \quad (17.21)$$

This effective charge can be shown to give exactly the strength of interaction with single electrons derived by Kim and colleagues.

These results are valid only for long wavelengths where the equations of motion and dielectric functions are independent of wave vector. They are often suitable for interpreting the data associated with Raman scattering.

17.3 Beyond the Long-Wavelength Approximation

Proceeding beyond the long-wavelength approximation introduces two new features, namely, the quantum dielectric function for electrons and the lattice dispersion of the optical modes. In bulk material, we can expand the perturbing potential and the disturbance of the electron density in 3D Fourier series:

$$V(\mathbf{r}, t) = \frac{\Omega}{(2\pi)^3} \int V_\mathbf{q}(t) e^{i\mathbf{q}\cdot\mathbf{r}} d\mathbf{q}$$

$$\delta n(\mathbf{r}, t) = \frac{\Omega}{(2\pi)^3} \int \delta n_\mathbf{q}(t) e^{i\mathbf{q}\cdot\mathbf{r}} d\mathbf{q} \quad (17.22)$$

The perturbing potential is the sum of the unscreened potential and the screened potential. The expectation response of the system is of the form

$$\langle \alpha_g | V_\mathbf{q}(t) | \alpha_h \rangle = \langle \alpha_g | V_{0\mathbf{q}}(t) | \alpha_h \rangle + \langle \alpha_g | V_{s\mathbf{q}}(t) | \alpha_h \rangle \tag{17.23}$$

where V_{0q} refers to the unscreened perturbing potential, V_{sq} refers to the screening potential produced by the action of the perturbing potential on the electron gas, and V_q is the self-consistent perturbing potential: the result of the original and screened potentials. V_{sq} must be proportional to the overall potential V_q viz: $V_{sq} = \chi V_q$, where χ is the susceptibility, and so

$$\langle \alpha_g | V_\mathbf{q}(t) | \alpha_h \rangle = \frac{\langle \alpha_g | V_{0\mathbf{q}}(t) | \alpha_h \rangle}{1 - \chi} \tag{17.24}$$

The denominator is the dielectric function

$$\frac{\varepsilon(\omega, q)}{\varepsilon_0} = 1 - \chi \tag{17.25}$$

The expectation value of the screening potential has the form

$$\langle \alpha_g | V_{s\mathbf{q}}(t) | \alpha_h \rangle = \frac{e^2}{(2\pi)^3 \varepsilon_0} \int \frac{1}{q^2} \sum_{\alpha_i \alpha_j} \Pi_{\alpha_i \alpha_j} G_{ij}^{gh}(q) \langle \alpha_i | V_\mathbf{q}(t) | \alpha_j \rangle d\mathbf{q}$$

$$\Pi_{\alpha_i \alpha_j} = \frac{f(E_{\alpha_j}) - f(E_{\alpha_i})}{E_{\alpha_j} - E_{\alpha_i} - \hbar\omega - i\hbar\delta} \tag{17.26}$$

$$G_{ij}^{gh}(q) = \iint \psi_g(\mathbf{r})\psi_h(\mathbf{r})\psi_i(\mathbf{r'})\psi_j(\mathbf{r'}) d\mathbf{r} d\mathbf{r'}$$

In bulk material, the wave function of the electron in a conduction band can be taken to be

$$\psi(\mathbf{k}) = \frac{1}{V^{1/2}} e^{i\mathbf{k}\cdot\mathbf{r}} \tag{17.27}$$

The G factor is then (normal processes only)

$$G_{ij}^{gh}(\mathbf{q}) = \delta_{\mathbf{k}_g, \mathbf{k}_h + \mathbf{q}} \delta_{\mathbf{k}_i, \mathbf{k}_j + \mathbf{q}} \tag{17.28}$$

The dielectric function for the electrons is the Lindhard expression:

$$\varepsilon_e(\omega, q) = -\frac{e^2}{Vq^2} \sum_\mathbf{k} \frac{f(E_{\mathbf{k}+\mathbf{q}}) - f(E_\mathbf{k})}{E_{\mathbf{k}+\mathbf{q}} - E_\mathbf{k} - \hbar\omega - i\hbar\delta} \tag{17.29}$$

158   Coupled Modes

Here, V is the volume of the cavity, q is the wave vector of the coupled mode, k is the wave vector of the electron, E_k is the energy, and $f(E_k)$ is the distribution function.

For simplicity, we will neglect the effect of other scattering mechanisms that may be present in addition to coupled-mode effects (otherwise we would have to use the Lindhard–Mermin expression).

The long-wavelength model that we have been considering takes no account of the Landau damping of plasma oscillations defined by frequency and wave vector (ω, q). When (ω, q) lies in the region where single-particle excitations can occur, the plasma mode becomes ill-defined, and we speak of fluctuations over a range of frequency. Considering the absorption of plasma energy to excite an electron with wave vector k to a new state with wave vector k', with $E(k') = E(k) + \hbar\omega$, implies that $\hbar\omega = (\hbar^2/2m^*)(q^2 + 2kq\cos\theta)$, giving the width of the single-particle excitation regime determined by $0 < k < 2k$. Thus, in the long-wavelength case, damping is negligible, and the coupled modes are well defined. Elsewhere there is a spectrum $0 < \omega < \infty$, within which the damped coupled modes continue to be defined by frequencies at which the total dielectric function vanishes (Fig. 17.1a):

$$\mathrm{Re}[\varepsilon_e(\omega, q) + \varepsilon_{LO}(\omega, q)] = 0 \qquad (17.30)$$

The lattice dispersion, assumed to be quadratic, enters the expression for the LO dielectric function:

$$\varepsilon_{LO}(\omega, q) = \varepsilon_\infty \frac{\omega^2 - \omega_L^2 + v_L^2 q^2}{\omega^2 - \omega_T^2 + v_L^2 q^2} \qquad (17.31)$$

where v_L is the velocity of the LA modes. The strength of interaction with single electrons is given by the fluctuation-dissipation theorem as being proportional to

$$\mathrm{Im}\left(-\frac{1}{\varepsilon(\omega, q)}\right) \qquad (17.32)$$

The total rate requires an integration over all ω. Rather than follow this procedure, it is intuitively more transparent to retain the concept of well-defined coupled modes. This allows us to make a bridge to the long-wavelength regime that preserves the ideas of phonon content and coupling-strength ratios.

The phonon content is no longer given by equation (17.19), since it was derived assuming the classical equation of motion for the electron and the classical expression for the dielectric function (equation (17.12)). As we noted, the Lindhard function reveals that the latter is approximately valid only for very long

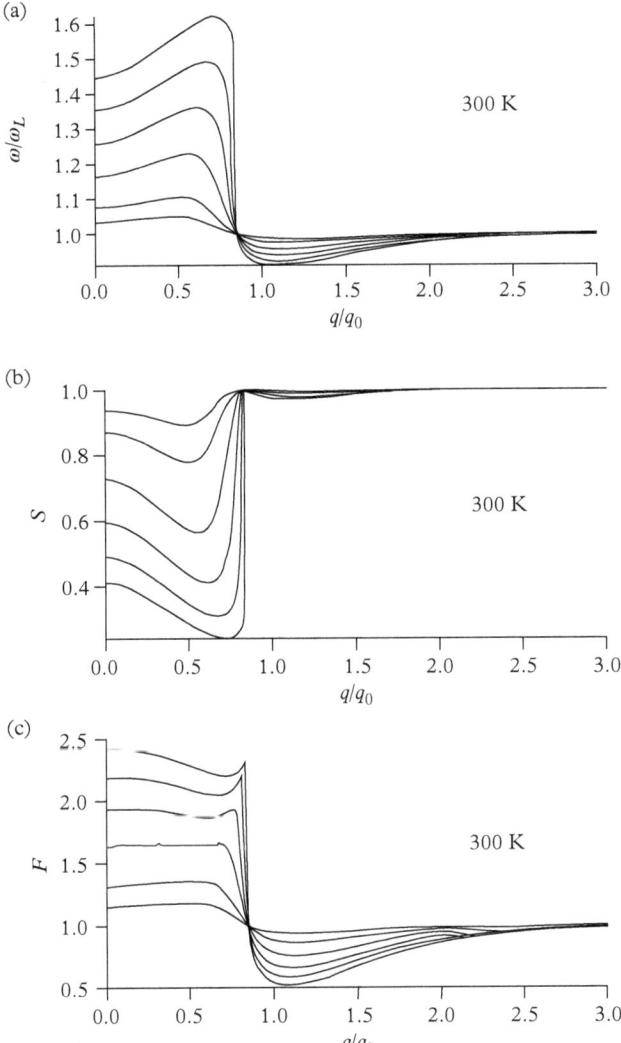

Figure 17.1 (a) Dependence of frequency of the coupled mode in GaN on wave vector for an electron temperature of 300 K. The electron densities from bottom up are 1, 2, 4, 6, 8, and 10×10^{18} cm^{-3}. (b) Dependence of the phonon content of the phonon-like coupled mode in GaN on wave vector for an electron temperature of 300 K. (c) Dependence of the ratio of screened to unscreened interaction strengths in GaN on wave vector for an electron temperature of 300 K.

wavelengths. However, if we abandon that assumption but retain the definition of the electron dielectric function through equation (17.29), we can relate the electron displacement to the optical displacement entirely in terms of the dielectric functions, using $\mathbf{u}_e = -(\varepsilon_e/ne)\mathbf{E}$ and $\mathbf{E} = (e^*/[V_0(\varepsilon_{LO} - \varepsilon_\infty)])\mathbf{u}_L$:

$$\mathbf{u}_e = -\frac{e^*}{neV_0} \frac{\mathrm{Re}\varepsilon_e(\omega, q)}{[\varepsilon_{LO}(\omega, q) - \varepsilon_\infty]} \mathbf{u}_L \qquad (17.33)$$

From the energy equation (equation (17.6)) we obtain

$$S = \left(1 + \frac{\omega_L^2 - \omega_T^2}{\omega_p^2}\left[\frac{\mathrm{Re}\varepsilon_e(\omega, q)}{[\varepsilon_{LO}(\omega, q) - \varepsilon_\infty]}\right]^2\right)^{-1} \qquad (17.34)$$

The dielectric function for electrons can be replaced by that for the LO mode using $\mathrm{Re}\varepsilon_e(\omega, q) + \varepsilon_{LO}(\omega, q) = 0$, and we obtain a general expression for the phonon content, using equation (17.31), as a function of frequency (Fig. 17.1b):

$$S = \left(1 + \frac{(\omega^2 - \omega_L^2 + v_L^2 q^2)^2}{\omega_p^2(\omega_L^2 - \omega_T^2)}\right)^{-1} \qquad (17.35)$$

We could have derived exactly this expression within the long-wavelength approximation in the same way. Given the variation of frequency with the plasma frequency in the long-wavelength approximation (equation (17.18)), the two expressions, equation (17.9) and equation (17.35), are identical. Equation (17.35) is the general expression for the phonon content with the frequency dependence on electron density and wave vector determined by equation (17.30). The ratio of the coupling strengths of screened to unscreened phonons (Fig. 17.1c) can be written as follows:

$$F = \left(\frac{\omega^2 - \omega_T^2 + v_L^2 q^2}{\omega_L^2 - \omega_T^2}\right)^2 \frac{\omega_L}{\omega} S \qquad (17.36)$$

The frequency that appears in equations (17.35) and (17.36) is determined by the satisfaction of the condition expressed by equation (17.30).

The temperature dependence of the dielectric function requires numerical work. However, analytic forms can be obtained for the extreme degenerate case and for the extreme non-degenerate case. It is useful to consider the equivalent expression to equation (17.29), namely:

$$\varepsilon_e(\omega) = \frac{e^2}{Vq^2} \sum_k f(E_k) \left[\frac{1}{E_{k-q} - E_k + \hbar\omega} + \frac{1}{E_{k+q} - E_k - \hbar\omega} \right] \quad (17.37)$$

17.3.1 Degenerate Case

For the case of a degenerate electron gas at $T = 0$, the dielectric function is

$$\varepsilon_e(\omega) = \frac{e^2 N(E_F)}{2q^2} \left[1 - \frac{1}{4\eta^3} \left(\begin{array}{l} [\gamma^2 - (1-2\gamma)\eta^2 + \eta^4] \ln \left| \frac{\eta + \eta^2 + \gamma}{\eta - \eta^2 - \gamma} \right| + \\ [\gamma^2 - (1+2\gamma)\eta^2 + \eta^4] \ln \left| \frac{\eta + \eta^2 - \gamma}{\eta - \eta^2 + \gamma} \right| \end{array} \right) \right] \quad (17.38)$$

where $N(E_F)$ is the density of states at the Fermi level, $\eta = q/2k_F$, $\gamma = \hbar\omega/4E_F$, and k_F is the Fermi wave vector.

17.3.2 Non-degenerate case

The distribution function is the Maxwell–Boltzmann form, and the dielectric function takes the form

$$\varepsilon_e(\omega) = -\frac{e^2 m^*}{4\pi^2 \hbar q^3 \lambda} \pi^{1/2} [Z(A_+) - Z(A_-)]$$

$$Z(A) = \frac{1}{\pi^{1/2}} \int_{-\infty}^{\infty} \frac{e^{-t^2}}{t - A} dt \quad (17.39)$$

$$A_\pm = \lambda^{1/2} \left(\frac{m^*\omega}{\hbar q} \pm \frac{q}{2} + i\delta \right)$$

where $\lambda = \hbar^2/2m^* k_B T$, k_B is Boltzmann's constant. $Z(A)$ is the plasma dispersion function which has the approximate form (Lowe and Barker 1985):

$$Z(A) = \frac{i\pi^{1/2} + (\pi - 2)A}{1 - i\pi^{1/2}A - (\pi - 2)A^2} \quad (17.40)$$

Figures 17.1 and 17.2 depict the dependence on wave vector of frequency, phonon content, and interaction strengths for GaN at 300 K and 600 K (lattice dispersion has been ignored). Anti-screening occurs at low wave vectors and screening at large wave vectors. Beyond the sharp drop in frequency with wave vector, plasma modes become ill defined as a consequence of Landau damping caused by single-particle excitation, and the phonon-like coupled mode reverts eventually to pure LO. The plasmon-like mode (not depicted here) disappears.

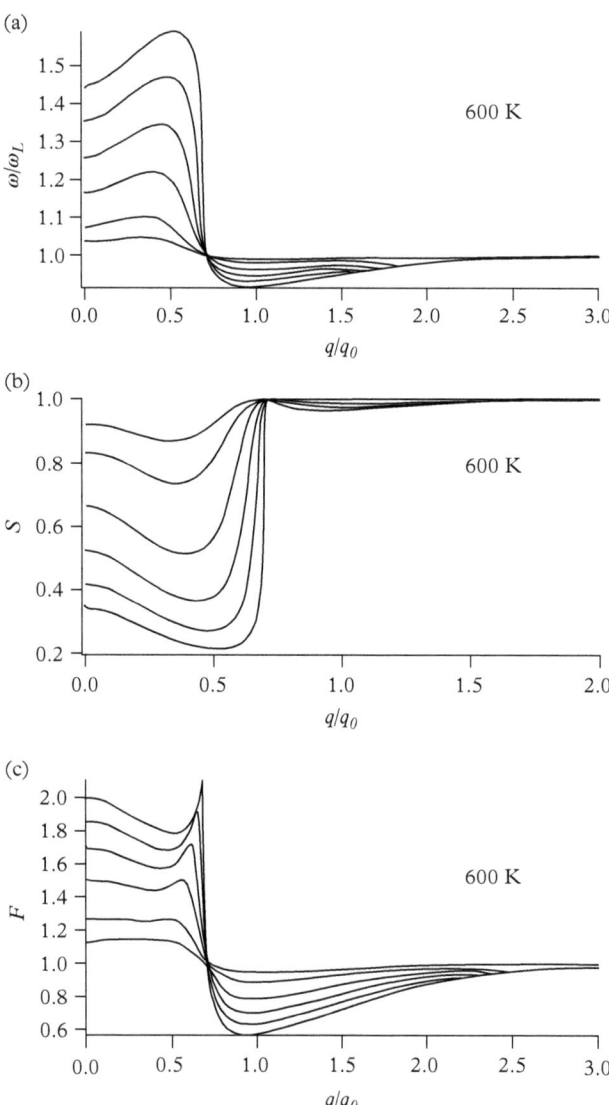

Figure 17.2 (a) Dependence of frequency of the coupled mode in GaN on wave vector for an electron temperature of 600 K. The electron densities from bottom up are 1, 2, 4, 6, 8, and 10 × 10^{18} cm^{-3}. (b) Dependence of the phonon content of the phonon-like coupled mode in GaN on wave vector for an electron temperature of 600 K. (c) Dependence of the ratio of screened to unscreened interaction strengths in GaN on wave vector for an electron temperature of 600 K.

17.4 Screening in Quasi-2D Structures

In quasi-2D nanostructures, the expansion of the perturbing potential has to be made in a 2D Fourier series:

$$V(\mathbf{r},t) = \frac{\sigma}{(2\pi)^2} \int V_\mathbf{q}(t) e^{i\mathbf{q}_\sigma \cdot \mathbf{r}} d\mathbf{q}_\sigma \tag{17.41}$$

where σ is the area and $\mathbf{q}_\sigma$ is the in-plane wave vector. The electron wave function is now

$$\psi(\mathbf{r}) = \frac{1}{\sigma^{1/2}} e^{i\mathbf{k}_\sigma \cdot \mathbf{r}} \psi(z) \tag{17.42}$$

Here, z is the direction in which there is some degree of confinement. The question of confinement also arises for the perturbing potential.

If it is not affected by confinement, such as the case for bulk phonons in a single heterostructure, the integral over q_z in the expression for V_s can be carried out to convert the 3D expansion to a 2D one using the standard integral

$$\int_{-\infty}^{\infty} \frac{e^{iq_z(z-z')}}{q_\sigma^2 + q_z^2} dq_z = \frac{\pi}{q_\sigma} e^{-q_\sigma |z-z'|} \tag{17.43}$$

The G factor becomes

$$G_{ij}^{gh} \rightarrow G_{n_1 n_2}^{n_3 n_4} = \delta_{\mathbf{k}_3, \mathbf{k}_4 + \mathbf{q}_\sigma} \delta_{\mathbf{k}_1, \mathbf{k}_2 + \mathbf{q}_\sigma} F_{n_1 n_2}^{n_3 n_4}(\mathbf{q}_\sigma) \tag{17.44}$$

where n labels the subbands. The form factor is

$$F_{n_1 n_2}^{n_3 n_4}(q_\sigma) \rightarrow F_{12}^{34}(q_\sigma) = \iint \psi_4(z)\psi_3(z)\psi_2(z')\psi_1(z') e^{-q_\sigma |z-z'|} dz dz' \tag{17.45}$$

The screening potential is

$$\langle 4 | V_{sq}(t) | 3 \rangle = \frac{e^2}{2\varepsilon_0 \sigma q_\sigma} \sum_{12\mathbf{k}} \Pi_{12} F_{12}^{34}(q_\sigma) \langle 2 | V_\mathbf{q}(t) | 1 \rangle \tag{17.46}$$

$$\Pi_{12} = \frac{f(E_{2,\mathbf{k}+\mathbf{q}_\sigma}) - f(E_{1,\mathbf{k}})}{E_{2,\mathbf{k}+\mathbf{q}_\sigma} - E_{1,\mathbf{k}} - \hbar\omega - i\hbar\delta}$$

A dielectric function can be defined in the case of a particular electron transition, subband n_3 to n_4, say, screened by electrons in subbands n_1 and n_2:

$$\varepsilon_{n_1 n_2}^{n_3 n_4} = \varepsilon_L \delta_{n_2 n_4} \delta_{n_1 n_3} - \frac{e^2}{2q\sigma} \sum_k \frac{f(E_{n_2 k+q}) - f(E_{n_1 k})}{E_{n_2 k+q} - E_{n_1 k} - \hbar\omega - i\hbar\eta} F_{n_2 n_1}^{n_4 n_3}(q) \qquad (17.47)$$

Because the screening potential depends on particular transitions, a straightforward dielectric function cannot be defined simply, except in certain cases. The simplest situation is when only the lowest subband is occupied, in which case the dielectric function becomes for intra-subband transitions:

$$\frac{\varepsilon_{11}^{11}(q_\sigma, \omega)}{\varepsilon_0} = 1 - \frac{e^2}{2\varepsilon_0 \sigma q_\sigma} \sum_k \Pi_{11} F_{11}^{11}(q_\sigma)$$

$$\Pi_{11} = \frac{f(E_{1,k+q_\sigma}) - f(E_{1,k})}{E_{1,k+q_\sigma} - E_{1,k} - \hbar\omega - i\hbar\delta} \qquad (17.48)$$

Assuming that the electrons are confined in a Fang–Howard state,

$$\psi(z) = \left(\frac{b^3}{2}\right)^{1/2} z e^{-bz/2}, \quad b = \left(\frac{33 m^* e^2 n_\sigma}{8\varepsilon_s \hbar^2}\right)^{1/3} \qquad (17.49)$$

and that the perturbing potential is in the form of a bulk travelling wave, the form factor is (see Fig. 17.3)

$$F_{11}^{11}(q_\sigma) = \frac{b}{8(q_\sigma + b)^3}(8b^2 + 9 q_\sigma b + 3 q_\sigma^2) \qquad (17.50)$$

Another 2D system is the quantum well. In this case, the electron is confined by the barriers, which, if taken to be infinitely high, results in a wave function for the ground state of the form

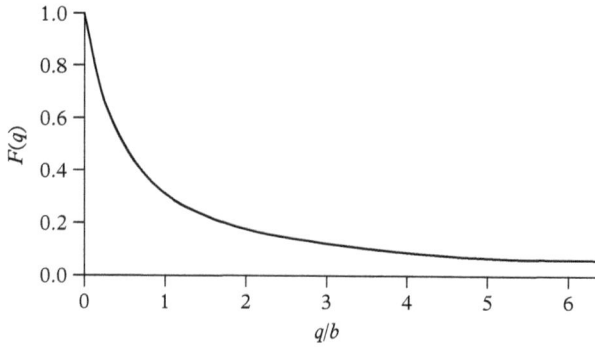

Figure 17.3 *Form factor associated with bulk phonons in a single heterostructure.*

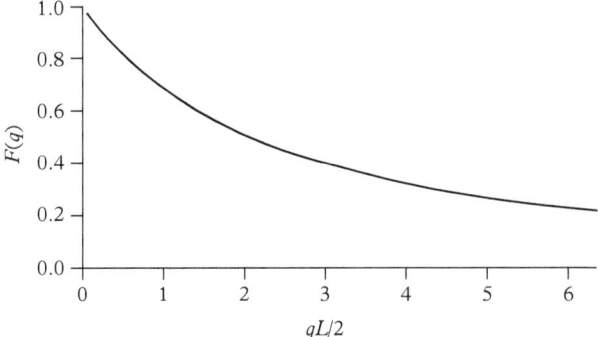

Figure 17.4 *Form factor associated with bulk phonons in a quantum well.*

$$\psi(z) = \left(\frac{2}{L}\right)^{1/2} \sin kz, \quad kL = n\pi \qquad (17.51)$$

Here, L is the well width. The form factor for the interaction within the ground state with a bulk-like travelling-wave potential is (see Fig. 17.4)

$$F_{11}^{11}(q_\sigma) = \frac{1}{2} \frac{[\eta(\eta^2 + \pi^2)(3\eta^2 + 2\pi^2) - \pi^4(1 - e^{-2\eta})]}{[\eta(\eta^2 + \pi^2)]^2} \qquad (17.52)$$

$$\eta = q_\sigma L/2$$

For brevity, we drop the subscript on q. Manipulation of equation (17.48) in the usual way gives

$$\varepsilon_e = \frac{e^2}{2q\sigma} F_{11}^{11}(q) \sum_k f(E_k) \left(\frac{1}{E_{k+q} - E_k - \hbar\omega - i\hbar\eta} + \frac{1}{E_{k-q} - E_k + \hbar\omega + i\hbar\eta} \right) \qquad (17.53)$$

Converting the sum to an integral, we get, for the real and imaginary parts ($\varepsilon_e = \varepsilon_{e1} + i\varepsilon_{e2}$):

$$\varepsilon_{e1} = \frac{e^2}{2q\sigma} F_{11}^{11}(q) \int f(E_k) \left(\frac{1}{E_{k+q} - E_k - \hbar\omega} + \frac{1}{E_{k-q} - E_k + \hbar\omega} \right) 2k\,dk\,d\theta\sigma/(2\pi)^2$$

$$\varepsilon_{e2} = \frac{\pi e^2}{2q\sigma} F_{11}^{11}(q) \int f(E_k) 2k\,dk\,d\theta\sigma/(2\pi)^2 [\delta(E_{k+q} - E_k - \hbar\omega) + \delta(E_{k-q} - E_k + \hbar\omega)]$$

$$(17.54)$$

Assuming a parabolic conduction band and introducing the conservation of momentum in the plane, we get

$$\varepsilon_{e1} = \frac{e^2 m^*}{(2\pi)^2 \hbar^2 q^2} F(q) \int_0^\infty f(E_k) dk \int_0^{2\pi} \left(\frac{1}{\cos\theta + (\alpha_+/k)} + \frac{1}{\cos\theta - (\alpha_-/k)} \right) d\theta \qquad (17.55)$$

where θ is the angle between k and q, and

$$\alpha_\pm = \frac{m^*\omega}{\hbar q} \pm \frac{q}{2} \qquad (17.56)$$

Carrying out the integration over angle, we get, with $\beta = (k_B T)^{-1}$ and $\lambda = \hbar^2 \beta / 2m^*$:

$$\varepsilon_{e1} = \begin{cases} \frac{e^2 m^*}{4\pi \hbar^2 q^2} F(q) \int_0^{\alpha_\pm^2} \frac{1}{1+e^{\lambda k^2 - \beta E_F}} \left(\frac{\operatorname{sgn}\alpha_+}{\sqrt{\alpha_+^2 - k^2}} - \frac{\operatorname{sgn}\alpha_-}{\sqrt{\alpha_-^2 - k^2}} \right) dk^2 & \alpha_\pm > k \\ 0 & \alpha_\pm < k \end{cases} \qquad (17.57)$$

Note that we have used a simplifying notation in which the upper limit of the integral α_+^2 refers to the first term in the bracket, and α_-^2 to the second.

For the imaginary part, we can put $\int_0^{2\pi} d\theta = 2 \int_{-1}^{1} d(\cos\theta)/\sin\theta$ and obtain

$$\varepsilon_{e2} = \begin{cases} \frac{e^2 m^*}{4\pi \hbar^2 q^2} F(q) \int_0^{\alpha_\pm} \frac{1}{1+e^{\lambda k^2 - \beta E_F}} \left(\frac{\operatorname{sgn}\alpha_+}{\sqrt{k^2 - \alpha_+^2}} - \frac{\operatorname{sgn}\alpha_-}{\sqrt{k^2 - \alpha_-^2}} \right) dk^2 & \alpha_\pm < k \\ 0 & \alpha_\pm > k \end{cases} \qquad (17.58)$$

Further analytic progress depends on whether the electron gas is degenerate or non-degenerate. The surface density of electrons is given by

$$N = \int_0^\infty 2k\, dk\, d\theta / (2\pi)^2 (1 + e^{\beta(E-E_F)})^{-1} = \frac{m^*}{\pi \hbar^2 \beta} \ln(1 + e^{\beta E_F}) \qquad (17.59)$$

17.4.1 Degenerate case

For $T \to 0$ and $E_F > 0$ equations (17.57) and (17.58) become

$$\varepsilon_{e1} = \frac{e^2 m^*}{4\pi \hbar^2 q^2} F(q) \int_0^{k_F^2} \left(\frac{\text{sgn}\,\alpha_+}{\sqrt{\alpha_+^2 - k^2}} - \frac{\text{sgn}\,\alpha_-}{\sqrt{\alpha_-^2 - k^2}} \right) dk^2$$

(17.60)

$$\varepsilon_{e2} = \frac{e^2 m^*}{4\pi \hbar^2 q^2} F(q) \int_{\alpha_\pm^2}^{k_F^2} \left(\frac{\text{sgn}\,\alpha_+}{\sqrt{k^2 - \alpha_+^2}} - \frac{\text{sgn}\,\alpha_-}{\sqrt{k^2 - \alpha_-^2}} \right) dk^2$$

Carrying out the integrations, and noting that $N = k_F^2/2\pi$, we get

$$\varepsilon_{e1} = \frac{Ne^2 m^*}{\hbar^2 q^2 k_F} F(q) \left(\frac{q}{k_F} - \text{sgn}\,\alpha_+[(\alpha_+/k_F)^2 - 1]^{1/2} + \text{sgn}\,\alpha_-[(\alpha_-/k_F)^2 - 1]^{1/2} \right)$$

$$\varepsilon_{e2} = \frac{Ne^2 m^*}{\hbar^2 q^2 k_F} F(q) \left([1 - (\alpha_+/k_F)^2]^{1/2} - [1 - (\alpha_-/k_F)^2]^{1/2} \right)$$

(17.61)

17.4.2 Non-degenerate case

Equation (17.57) becomes

$$\varepsilon_{e1} = \frac{e^2 m^* e^{\beta E_F}}{4\pi \hbar^2 q^2} F(q) \int_0^{\alpha_\pm^2} e^{-\lambda k^2} \left(\frac{\text{sgn}\,\alpha_+}{\sqrt{\alpha_+^2 - k^2}} - \frac{\text{sgn}\,\alpha_-}{\sqrt{\alpha_-^2 - k^2}} \right) dk^2 \quad (17.62)$$

The integral can be expressed in terms of incomplete gamma functions:

$$\varepsilon_{e1} = \frac{Ne^2 \beta}{2q^2} F(q)(\text{sgn}\,\alpha_+ I_+ - \text{sgn}\,\alpha_- I_-) \quad (17.63)$$

$$I_\pm = (-\lambda)^{1/2} e^{-\alpha_\pm^2 \lambda} \gamma(1/2, -\alpha_\pm^2 \lambda) \quad (17.64)$$

Alternatively, the incomplete gamma function can be expressed in terms of the confluent hypergeometric function:

$$\gamma(\rho, x) = \rho^{-1} x^\rho e^{-x} {}_1F_1(1; 1 + \rho; x) \quad (17.65)$$

giving

$$\varepsilon_{e1} = \frac{Ne^2 \beta}{2q^2} F(q) [\alpha_+ {}_1F_1(1; 3/2; -\alpha_+^2 \lambda) - \alpha_- {}_1F_1(1; 3/2; -\alpha_-^2 \lambda)] \quad (17.66)$$

Noting that $N = m^* e^{\beta E_F}/\pi \hbar^2 \beta$, this is the result obtained by Lee and Galbraith (1999).

A result in the same form as in the case of a degenerate distribution can be obtained by integrating equation (17.62) by parts. We get

$$\varepsilon_{e1} = \frac{Ne^2\beta}{2q^2} F(q) \left(q + \lambda \int_0^{\alpha_\pm^2} e^{-\lambda k^2} [-\text{sgn}\,\alpha_+ (\alpha_+^2 - k^2)^{1/2} + \text{sgn}\,\alpha_- (\alpha_-^2 - k^2)^{1/2}] dk^2 \right) \quad (17.67a)$$

whence

$$\varepsilon_{e1} = \frac{Ne^2\beta}{2q^2} F(q) [q - (2/3)\lambda [\alpha_{+1} F_1(1; 5/2; -\alpha_+^2 \lambda) - \alpha_{-1} F_1(1; 5/2; -\alpha_-^2 \lambda)] \quad (17.67b)$$

The imaginary part is straightforward:

$$\varepsilon_{e2} = \frac{e^2 m^* e^{\beta E_F}}{4\pi \hbar^2 q^2} F(q) \int_{\alpha_\pm^2}^\infty e^{-\lambda k^2} \left(\frac{1}{\sqrt{k^2 - \alpha_-^2}} - \frac{1}{\sqrt{k^2 - \alpha_+^2}} \right) dk^2 \quad (17.68)$$

Integration by parts gives

$$\varepsilon_{e2} = \frac{Ne^2\beta}{2q^2} F(q) \frac{\Gamma(3/2)}{\lambda^{1/2}} \left(e^{-\alpha_-^2 \lambda} - e^{-\alpha_+^2 \lambda} \right) \quad (17.69)$$

(Note: $\Gamma(3/2) = \pi^{1/2}/2$).

17.4.3 Long-Wavelength Solutions

With $q \sim 0$ we have $\alpha_\pm^2 \to \infty$ and $F(q) \sim 1$. For the degenerate gas, the roots in equation (17.61) have to be expanded up to q^3. When this is done we obtain

$$\varepsilon_e = -\varepsilon_\infty \frac{\omega_p^2}{\omega^2}$$

$$\omega_p^2 = \frac{Ne^2 q F(q)}{2\varepsilon_\infty m^*} \quad (17.70)$$

The plasma frequency in 2D depends on the root of the wave vector.

In the non-degenerate case, $_1F_1(1;3/2;z) \xrightarrow{|z|\to\infty} -1/(2z)$, and

$$\varepsilon_e = \frac{Ne^2\beta}{4q^2\lambda}F(q)\left(\frac{\alpha_- - \alpha_+}{\alpha_+\alpha_-}\right) = -\varepsilon_\infty \frac{\omega_p^2}{\omega^2 - \omega_q^2} \qquad (17.71)$$

$$\omega_q^2 = \frac{\hbar^2 q^4}{4m^{*2}}$$

The lattice contribution (including the lattice dispersion) is

$$\varepsilon_{LO} = \varepsilon_\infty \frac{\omega^2 - \omega_L^2 + i\eta\omega + v_L^2(q^2 + q_z^2)}{\omega^2 - \omega_T^2 + i\eta\omega + v_L^2(q^2 + q_z^2)} \qquad (17.72)$$

where η is the reciprocal lifetime of the optical phonon.

Taking $\eta = 0$ and neglecting dispersion, we get for the coupled-mode frequencies, defined by $\varepsilon = \varepsilon_{LO} + \varepsilon_e = 0$:

$$\begin{aligned}\omega_+^2 &= \omega_L^2 + \omega_p^2 + \omega_q^2 \\ \omega_-^2 &= \omega_{ps}^2 + \omega_q^2\end{aligned} \qquad (17.73)$$

$$\omega_{ps}^2 = \omega_p^2 \varepsilon_\infty / \varepsilon_s \qquad (17.74)$$

The two branches are depicted in Figure 17.5.

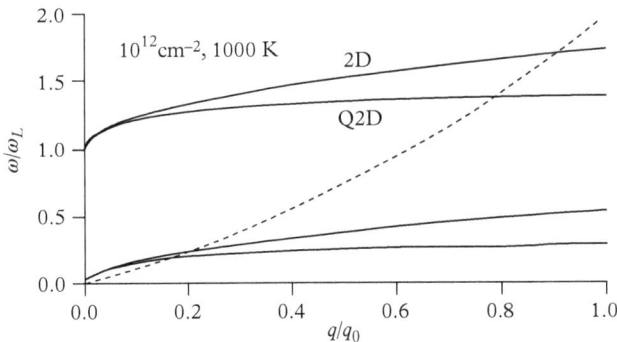

Figure 17.5 *Coupled-mode frequencies in a GaN single heterostructure for strict 2D case and quasi-2D case (form factor included). The areal electron density is 10^{12} cm^{-2}. The dashed line is the boundary of the single-electron excitation spectrum for $k = \sqrt{2m^*k_BT/\hbar^2}$, $T = 1000$ K.*

17.5 Coupling to Hybrids

As far as the interaction of electrons with polar optical phonons is concerned, these estimates of the effect of screening in single heterostructures and quantum wells can only be very approximate, being derived under the assumption that the LO phonons are normal modes of the structure and entirely unconfined as they are in the bulk. In fact, they are neither.

The LO hybrid modes we consider are those having an LO component, IF component, and a TO component. Plasma coupling has no effect on pure transverse waves modes. The latter's electric displacement satisfies $\nabla \cdot \mathbf{D} = 0$, so it satisfies Gauss's equation independently of the dielectric function. The polaronic origin of the IF mode defines it as an electromagnetic wave and therefore transversely polarized. The field associated with the IF mode remains unscreened. We just have to consider the effect on the LO/plasma components, which are longitudinally polarized. A coupled mode—plasma with hybrid—must therefore be described by a dielectric function whose real part must vanish so that $\nabla \cdot \mathbf{D} = \nabla(\varepsilon \mathbf{E}) = 0$. The dielectric function must now include the modification to the electron dielectric function due to the lower dimensionality of the nanostructure and the confinement of the electron gas in the nanostructure. As far as the author is aware, an account of coupled plasmon–hybridon modes has yet to be given.

The perturbing potential derives essentially from a hybrid of LO and IF modes, both of which can suffer the effects of confinement in the z direction. The potential is given in equation (13.1) for the single heterostructure, and in equation (14.4) for the quantum well, and its dependence on the wave vector of the perturbation is determined by energy normalization. Moreover, lattice dispersion distinguishes the interaction strength between the long wavelength, LO-like hybrid, and the short wavelength, IF-like hybrid. In the case of the single heterostructure, an integration over q_z can still be carried out. This leads to the following $G(z,z')$ factors:

$$G(z, z') = e^{-q|z'-z|} \quad \text{bulk}$$

$$G(z, z') = e^{-q|z'-z|} - e^{-q|z'+z|} \quad \text{LO-like} \tag{17.75}$$

$$G(z, z') = e^{-q|z'+z|} \quad \text{IF-like}$$

where q is the in-plane wavevector. However, the coupling between plasmon and phonon can occur only for wave vectors small enough to avoid heavy Landau damping.

In the case of the quantum well, integrations over q_z must be replaced by sums over allowed magnitudes.

Confinement in quantum wires implies that integrations over q_r have to be replaced by sums over the allowed values of the phonon vector. Where extreme accuracy is not required, it may seem acceptable that screening can be neglected entirely. This is sometimes justified by the argument that, in any case, there is screening and anti-screening, and we can take one to cancel out the other. Even if this desperate choice is ruled out, any calculation is bound to be extremely laboriously computer intensive, and therefore limited in generality, without other approximations being made, such as the assumption of $T = 0$ degeneracy for the electron gas, or the assumption of bulk phonons, ignoring confinement.

17.6 Quasi-1D Cylindrical Structures

In quasi-1D cylindrical nanostructures, the electron wave function for the ground state corresponding to complete confinement is

$$\psi(r) = \frac{\mathcal{J}_0(k_0 r)}{(\pi R^2)^{1/2} \mathcal{J}_1(k_0 R)} \quad \mathcal{J}_0(k_0 R) = 0 \quad (17.76)$$

The wave vector of the perturbing potential, taken once again to be an unconfined travelling wave, now has an axial component, q, and radial components q_r, thus, exploiting standard integrals:

$$\iint \frac{e^{i(q_r|\mathbf{r}-\mathbf{r}'|\cos\theta)}}{q^2 + q_r^2} q_r dq_r d\theta = 2\pi \int_0^\infty \frac{\mathcal{J}_0(q_r|\mathbf{r}-\mathbf{r}'|)}{q^2 + q_r^2} q_r dq_r = 2\pi K_0(q_r|\mathbf{r}-\mathbf{r}'|)$$

(17.77)

The G factor becomes

$$G_{n_1 n_2}^{n_3 n_4} = \delta_{k_3,k_4+q} \delta_{k_1,k_2+q} F_{n_1 n_2}^{n_3 n_4}$$

$$F_{n_1 n_2}^{n_3 n_4} = \iint \psi_4(\mathbf{r}) \psi_3(\mathbf{r}) \psi_2(\mathbf{r}') \psi_1(\mathbf{r}') K_0(q_r|\mathbf{r}-\mathbf{r}'|) d\mathbf{r} d\mathbf{r}'$$

(17.78)

When all electrons are in the lowest subband, the dielectric response function is

$$\frac{\varepsilon_{11}^{11}(q,\omega)}{\varepsilon_0} = 1 - \frac{e^2}{2\pi \varepsilon_0 L} \sum_k \Pi_{11} F_{11}^{11}(q)$$

$$\Pi_{11} = \frac{f(E_{1,k+q}) - f(E_{1,k})}{E_{1,k+q} - E_{1,k} - \hbar\omega - i\hbar\delta}$$

(17.79)

Furthermore

$$K_0(q|\mathbf{r}-\mathbf{r}'|) = K_0\left|(q(r^2 + r'^2 - 2rr'\cos\theta_{rr'})^{1/2}\right) = K_0(qr) I_0(qr') \quad (17.80)$$

In equation (17.51) for the form function, the integration over angle is straightforward:

$$\int_0^{2\pi} K_0(qr)I_0(qr)d\theta' = 2\pi \begin{cases} K_0(qr)I_0(qr') \big|_{r'<r} \\ K_0(qr')I_0(qr) \big|_{r<r'} \end{cases} \quad (17.81)$$

Analytic solutions for the form factor can be found only by making approximations to the electron wave functions (Lee and Spector 1983, 1985; Gold and Ghazali 1990). Bennett and colleagues (1994) have explored the dependence of the plasmon spectrum on the cross section of the wire—circular, elliptical, and rectangular.

17.7 Mobility

Solving the Boltzmann equation for the interaction between electrons and coupled modes involves computer-intensive numerical work, even when bulk LO modes are assumed. In the absence of screening, the frequency involved in the ladder technique is the LO frequency; in the absence of dispersion, a constant quantity. Dispersion and coupling make this frequency variable, which complicates the application of the ladder technique (Anderson D.R., (2003) Ph.D. Thesis, University of Essex). Replacing bulk LO modes with hybrids adds another level of complexity. The problem, as far as the author is aware, has yet to be tackled. We have to conclude that a reliable estimate of LO phonon-limited mobility in nanostructures has yet to be made.

18
Hot Phonon Lifetime

18.1 Introduction

It is well known that the performance of high-power semiconductor devices can be significantly degraded by the production of hot phonons. In III-V compounds at high electric fields, hot electrons interact strongly with polar optical phonons, with the result that the phonon population is driven far from thermodynamic equilibrium and can be described as being hot. The increase in phonon number results naturally in an increase of electrical resistance, which affects device performance. How strong this effect is depends on the lifetime of these hot phonons. In the harmonic approximation, lattice waves live forever. Models of phonon lifetime therefore focus on the anharmonicity of the lattice, the principal feature of which is to allow phonons to interact. In the simplest process (Type 1), a phonon is annihilated and two phonons are created; alternatively (Type 2), two phonons interact and are annihilated and a third created. All three phonons can be LA or TA modes or, when optical phonons are involved, one optical mode and two acoustic modes or two optical modes and one acoustic mode. Four-phonon and higher order processes are possible, but they can often be ignored relative to three-phonon processes. In all cases, crystal momentum and energy are conserved, a condition that rules out a number of interactions. The strength of the anharmonicity can be deduced from thermal expansion, frequency shifts, and measurements of lifetime.

The theory of phonon lifetime in bulk material is well understood. Its role in the production of hot phonons is immediately evident from the net rate at which phonons are produced by hot electrons:

$$\frac{dn(q,\omega)}{dt} = G(q,\omega) - \frac{n(q,\omega) - n(q,\omega,T_L)}{\tau_p(q,\omega)} \tag{18.1}$$

Here, $n(q,\omega)$ is the phonon occupation number, $G(q,\omega)$ is the net rate of emission with wave-vector q and frequency ω, $n(q,\omega,T_L)$ is the thermodynamic-equilibrium number at the lattice temperature, and $\tau_p(q)$ is the lifetime. The net emission rate is

$$G(q,\omega) = W_{em}(q,\omega)[n(q,\omega) + 1] - W_{abs}(q,\omega)n(q,\omega) \tag{18.2}$$

Hybrid Phonons in Nanostructures. First Edition. B.K. Ridley. © B.K. Ridley 2017.
Published in 2017 by Oxford University Press. DOI: 10.1093/acprof:oso/9780198788362.001.0001

The emission and absorption rates involve a sum over the electron distribution function. The distribution function for hot electrons can be regarded as non-degenerate with an electron temperature T_e. It is useful to consider the case when the phonon temperature is equal to the electron temperature, which would define thermodynamic equilibrium. In such a case, the net production of phonons would be zero by the principle of detailed balance, and

$$G(q,\omega,T_e) = W_{em}(q,\omega,T_e)[n(q,\omega,T_e)+1] - W_{abs}(q,\omega,T_e)n(q,\omega,T_e) = 0 \quad (18.3)$$

Thus

$$W_{abs}(q,\omega,T_e) = W_{em}(q,\omega,T_e)\frac{n(q,\omega,T_e)+1}{n(q,\omega,T_e)} \quad (18.4)$$

where $n(q,\omega,T_e)$ is the number at the temperature of the electron gas. Substitution of the absorption rate gives for the occupation number at the steady state:

$$n(q,\omega) = n(q,\omega,T_e)\frac{n(q,\omega,T_L) + \lambda(q,\omega,T_e)}{n(q,\omega,T_e) + \lambda(q,\omega,T_e)} \quad (18.5)$$

$$\lambda(q,\omega,T_e) = W_{em}(q,\omega,T_e)\tau_p(q,\omega) \quad (18.6)$$

Note that $W_{em}(q,\omega,T_e)$ is the total spontaneous emission rate for the mode with wave vector q. For a non-degenerate electron gas it is given by

$$W_{em}(q,\omega,T_e) = \frac{1}{2}W_0(q)\left(\frac{\hbar\omega(q)}{E_q^3}\right)^{1/2}\frac{N_e}{N_c}k_B T_e e^{-E_1/k_B T_e} \quad (18.7)$$

$$E_q = \frac{\hbar^2 q^2}{2m^*}, \quad E_1 = \frac{[\hbar\omega(q) + E_q]^2}{4E_q}$$

where N_c is the effective density of states (including spin degeneracy) and N_e is the electron density. The rate $W_0(q)$, in general, is modified by coupled-mode effects. It is related to the bare-phonon rate, W_0, as follows (equation (17.29)):

$$W_0(q) = W_0\left(\frac{\omega^2 - \omega_T^2}{\omega_L^2 - \omega_T^2}\right)^2\left(\frac{\omega_L}{\omega}\right)S \quad (18.8)$$

where S is the phonon content.

Figure 18.1 shows $W_{em}(q)$ for 300 K and 3000 K and for a range of densities. $W_{em}(q)$ defines the wave-vector range of excited modes, roughly between 0 and $3q_0$. There is the expected increase in rate with density, and there is the shift towards low q due to Landau damping. The increasing spikiness with density reflects the rapid transition from anti-screening to screening.

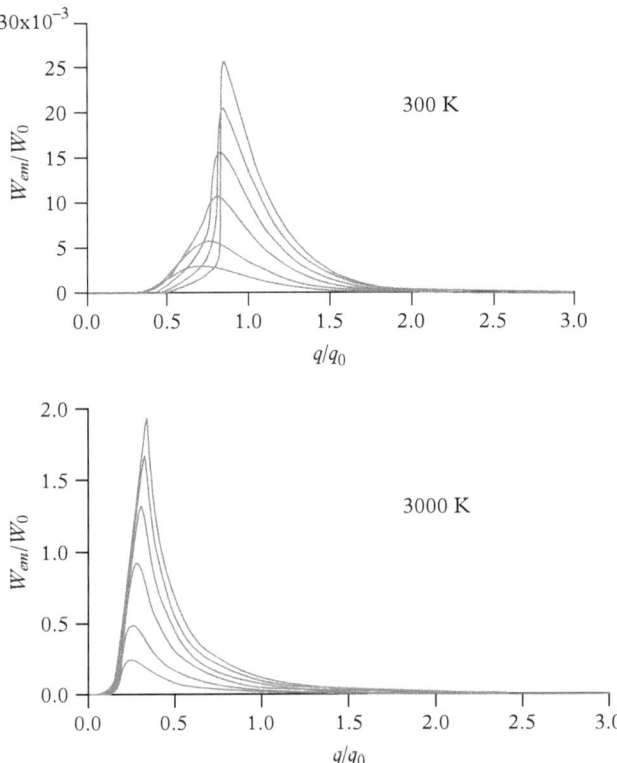

Figure 18.1 *Spontaneous emission rate for electron densities 1, 2, 4, 6, 8, 10 × 10^{18} cm^{-3}.*

18.2 Lifetime

The role of hybrid modes in the emission rate and in coupled modes has been discussed in previous chapters. There remains the question of the effect of hybridization on lifetime.

We limit attention to isotropic media and to three-phonon processes. The Hamiltonian for lattice vibrations in lowest order can be written as

$$H = \sum_{\mathbf{r},i,j} M_i^{1/2}\omega_i M_j^{1/2}\omega_j \mathbf{u_i}.\mathbf{u_j} \tag{18.9}$$

Here, $M_{i,j}$ are the appropriate masses in the unit cell (total for acoustic, relative for optical modes). Let us first review the case for bulk material. Expanding the displacement in a Fourier series enables us to write the displacement as an operator operating on the number state of the mode:

$$\mathbf{u} = \sum_q \left(\frac{\hbar}{2NM\omega}\right)^{1/2} [\hat{\mathbf{r}} a_q e^{i\mathbf{q}\cdot\mathbf{r}} + \hat{\mathbf{r}}^* a_{-q}^\dagger e^{-i\mathbf{q}\cdot\mathbf{r}}] \qquad (18.10)$$

Here, N is the number of unit cells, and $\hat{\mathbf{r}}$ is a unit vector. In terms of the annihilation and creation operators the unperturbed Hamiltonian is

$$H_0 = \sum_q \hbar\omega_q (a_q^\dagger a_q + 1/2) \qquad (18.11)$$

We take the anharmonic perturbation to be simply quantified by the change in frequency:

$$H_1 = \sum_{\mathbf{r},i,j} M_i^{1/2} \omega_i M_j^{1/2} \omega_j \left(\frac{\delta\omega_i}{\omega_i} + \frac{\delta\omega_j}{\omega_j}\right) \mathbf{u_i}.\mathbf{u_j} \qquad (18.12)$$

In an analogy with the deformation potential, the change of frequency of an acoustic mode can be taken to be proportional to the strain associated with an incident mode having an amplitude with magnitude u_k:

$$\frac{\delta\omega}{\omega} = iq_k u_k \qquad (18.13)$$

(Fuller accounts can be found in Ziman (1960) and Srivastarva (1990)). In the case of an optical mode:

$$\frac{\delta\omega}{\omega} = \Gamma u_k \qquad (18.14)$$

The mode that initiates the interaction is known as the promoting mode (Stoneham 1975). The perturbation Hamiltonian for three-phonon processes is then

$$H_1 = 2 \sum_{\mathbf{r},i,j} \Gamma M_i^{1/2} \omega_i M_j^{1/2} \omega_j u_k \mathbf{u_i}.\mathbf{u_j} \qquad (18.15)$$

with suitable choice of Γ. Inserting equation (18.10) leads to the result for bulk material:

$$H_1 = \left(\frac{2}{N^{3/2}}\right) \sum_{\mathbf{q}_k,\mathbf{q}_i,\mathbf{q}_j} \sum_{\mathbf{r},i,j} \Gamma M_i^{1/2} \omega_i M_j^{1/2} \omega_j \left(\frac{\hbar}{2M_k\omega_k}\right)^{1/2} \left(\frac{\hbar}{2M_i\omega_i}\right)^{1/2} \left(\frac{\hbar}{2M_j\omega_j}\right)^{1/2} \hat{A}$$

$$\hat{A} = (a_{\mathbf{q}_k} e^{iq_{xk}x} + a_{-\mathbf{q}_k}^\dagger e^{-iq_{xk}x})[\mathbf{r_i}.\mathbf{r_j}](a_{\mathbf{q}_i} e^{iq_{xi}x} + a_{-\mathbf{q}_i}^\dagger e^{-iq_{xi}x})(a_{\mathbf{q}_j} e^{iq_{xj}x} + a_{-\mathbf{q}_j}^\dagger e^{-iq_{xj}x})$$
$$(18.16)$$

The rate is given by the Fermi Golden Rule:

$$W = \frac{2\pi}{\hbar} \int |H_1|^2 \, \delta(\hbar\omega_k - \hbar\omega_i - \hbar\omega_j) \, dN_f \qquad (18.17)$$

The anharmonic interaction transforms mode A into daughter modes $B + C$ with conservation of energy. The sum over space ensures that the crystal momentum is constrained:

$$\mathbf{q}_k - \mathbf{q}_i - \mathbf{q}_j = \mathbf{g} \qquad (18.18)$$

where $\mathbf{g}$ is a vector of the reciprocal lattice. In normal processes, $\mathbf{g} = 0$.

Our immediate interest is the lifetime of an LO mode. In cubic semiconductors three possible decay routes have been identified:

$$\begin{aligned} &LO \rightarrow 2LA \text{ (Klemens 1966)} \\ &LO \rightarrow LO + LA \text{ or } TA \text{ (Vallée and Bogani 1991)} \qquad (18.19) \\ &LO \rightarrow TO + LA \text{ or } TA \text{ (Ridley 1996)} \end{aligned}$$

In GaAs quantum wells, the Klemens process has been shown to give the same result for the LO lifetime as for bulk (Usher and Srivastava 1994), indicating that the strength of the anharmonic interaction is not affected by confinement. Here we present a simple model that the anharmonic interaction is largely unaffected by hybridization, whatever the channel of decay.

In isotropic quantum wells, the displacement takes the form

$$\mathbf{u} = \left(\frac{1}{N}\right)^{1/2} \sum_{\mathbf{q}} A_\mathbf{q} e^{iq_x x} [\hat{\mathbf{x}} u_x(\mathbf{q}, z) + \hat{\mathbf{z}} u_z(\mathbf{q}, z)] \qquad (18.20)$$

Here, x is an arbitrary direction in the interface between barrier and well, $\hat{\mathbf{x}}, \hat{\mathbf{z}}$ are unit vectors, and $\mathbf{q} = (q_x, q_z)$. In the hybrid model, u_x and u_z are linear combinations of LO, TO, and IF modes. We adopt the simplifying assumption that the anharmonic strength is quantified by the magnitude of the displacement. Thus

$$\mathbf{u}_\mathbf{q} \rightarrow \left(A_\mathbf{q}^2 (|u_x|^2 + |u_z|^2)\right)^{1/2} \qquad (18.21)$$

Furthermore, we replace $u_{x,z}$ by their averages over z:

$$\mathbf{u}_\mathbf{q} \rightarrow \left(A_\mathbf{q}^2 \left(\langle |u_x|^2 \rangle + \langle |u_z|^2 \rangle\right)\right)^{1/2} \qquad (18.22)$$

Here, A_q is determined by energy normalization:

$$\left(A_q^2 \left(\langle|u_x|^2\rangle + \langle|u_z|^2\rangle\right)\right)^{1/2} = \left(\frac{\hbar}{2M\omega_q}\right)^{1/2} (a_q e^{iq_x x} + a^\dagger_{-q} e^{-iq_x x}) \quad (18.23)$$

Thus

$$H_1 = \left(\frac{2}{N^{3/2}}\right) \sum_{q_k, q_i, q_j} \sum_{r, i, j} \Gamma M_i^{1/2} \omega_i M_j^{1/2} \omega_j \left(\frac{\hbar}{2M_k \omega_k}\right)^{1/2} \left(\frac{\hbar}{2M_i \omega_i}\right)^{1/2} \left(\frac{\hbar}{2M_j \omega_j}\right)^{1/2} \hat{A}$$

$$\hat{A} = (a_{q_k} e^{iq_{xk} x} + a^\dagger_{-q_k} e^{-iq_{xk} x})[\mathbf{r}_i \cdot \mathbf{r}_j](a_{q_i} e^{iq_{xi} x} + a^\dagger_{-q_i} e^{-iq_{xi} x})(a_{q_j} e^{iq_{xj} x} + a^\dagger_{-q_j} e^{-iq_{xj} x}) \quad (18.24)$$

We obtain an expression that is identical to the bulk result. Hybridization has no effect.

Confinement, however, does have an effect, since q_z is restricted by the boundary conditions. Let us consider the Klemens process in a slab where the confinement in the z direction is maximized. In this, the LO mode is annihilated and two LA modes are created. Assume that the LO mode is the promoting mode:

$$\hat{A} = [n(\omega_k)]^{1/2}[n(\omega_i) + 1]^{1/2}[n(\omega_j) + 1]^{1/2} \quad (18.25)$$

To obtain the net rate of decay it is necessary to take into account the reverse process in which two LA modes annihilate and produce a LO mode. The net rate then involves the factor

$$\hat{A}^2_{net} = [n(\omega_k)][n(\omega_i) + 1][n(\omega_j) + 1] - [n(\omega_k) + 1][n(\omega_i)][n(\omega_j)]$$
$$= [n(\omega_k)][n(\omega_i) + n(\omega_j) + 1] - n(\omega_i)n(\omega_j) \quad (18.26)$$

At thermodynamic equilibrium the net rate must vanish, thus

$$[\bar{n}(\omega_k)][\bar{n}(\omega_i) + \bar{n}(\omega_j) + 1] - \bar{n}(\omega_i)\bar{n}(\omega_j) = 0 \quad (18.27)$$

The bar over the phonon number denotes equilibrium values. Assuming that the daughter modes remain substantially at equilibrium, we obtain the net rate proportional to

$$W \propto [n(\omega_L) - \bar{n}(\omega_L)][\bar{n}(\omega_i) + \bar{n}(\omega_j) + 1] \quad (18.28)$$

For normal processes,

$$q_{xk} - q_{xi} - q_{xj} = 0 \quad (18.29)$$

If θ_{ij} is the angle between q_{xi} and q_{xj},

$$q_{xk}^2 = q_{xi}^2 + q_{xj}^2 + 2q_{xi}q_{xj}\cos\theta_{ij} \tag{18.30}$$

Assume, further, that the LO mode is near the zone centre and dispersion can be ignored. The frequency for all q is ω_L. Near the zone centre we can, take $q_{xk} \sim 0$, in which case $q_{xi} \approx -q_{xj}$, $\cos\theta_{ij} \approx -1$, and therefore $\mathbf{r_i.r_j} = -1$. Conservation of energy implies

$$\omega_i = \omega_j = \omega_L/2 \tag{18.31}$$

Adopting the Debye model for acoustic modes, we take the frequency to be linearly dependent on wave vector through the velocity of sound, v_L:

$$\omega_i = v_L(q_x^2 + q_z^2)_i^{1/2} = \omega_L/2 \tag{18.32}$$

This condition implies that q_z and q_x are related:

$$q_z = \left(\frac{\omega_L^2}{2v_L^2} - q_x^2\right)^{1/2} \tag{18.33}$$

Symmetry considerations mean that the two LA modes have to have the same symmetry, that is, either symmetric or antisymmetric. The connection rules (see Chapter 6) define

$$\begin{cases} q_zL = (2m-1)\pi & \text{symmetric} \\ q_zL = 2m\pi & \text{antisymmetric} \end{cases} \tag{18.34}$$

where m is an integer. The sum over q_z gives a countable number of modes in each case that satisfies equation (18.32). The maximum numbers ($q_x = 0$) are

$$\begin{cases} m \leq \frac{1}{2}\left(\frac{\omega_L L}{2v_L\pi} + 1\right) & \text{symmetric} \\ m \leq \frac{\omega_L L}{4v_L\pi} & \text{antisymmetric} \end{cases} \tag{18.35}$$

Summing the contributions from the symmetric and antisymmetric solutions gives

$$m \leq \frac{1}{2}\left(\frac{\omega_L L}{v\pi} + 1\right) \tag{18.36}$$

A calculation of the net rate gives

$$W_{net} \leq \frac{\Gamma^2 \hbar \omega_L^2}{32\pi \rho_L v^2} \left(\frac{\omega_L L}{v\pi} + 1 \right) [2\bar{n}(\omega_L/2) + 1] \qquad (18.37)$$

Here, ρ_L is the reduced-mass density. This is to be compared to the bulk rate

$$W_{net} = \frac{\Gamma^2 \hbar \omega_L^3}{32\pi \rho_L v^3} [2\bar{n}(\omega_L/2) + 1] \qquad (18.38)$$

The ratio is

$$\frac{W_{conf}}{W_{bulk}} \leq \left(1 + \frac{v\pi}{\omega_L L} \right) \qquad (18.39)$$

The smallest dimension for which there is a confined LA in a GaAs slab ($m = 1$) is given by

$$L = 2\pi v_L / \omega_L \approx 0.5 nm \qquad (18.40)$$

This is a dimension far smaller than any continuum model of the nanostructure assumes. In general, we can conclude that the effect of confinement on the lifetime is small, but tending to lengthen the lifetime.

18.3 Thermal Conductivity

The lifetime enters the account of thermal conductivity. It is also interesting to ask what elements of hybridization affect the conduction of heat. The transfer of heat energy from hot to cold is effected by the flow of phonons, which, like electrons in electrical conduction, is obstructed by collisions with lattice imperfections, boundary roughness, and electrons, and collisions that tend to destroy the momentum of the phonon. But even in perfect insulating crystals, collisions occur with other phonons via the anharmonic interaction. Let us focus on this situation, since it is intrinsic to the perfect crystal.

At thermodynamic equilibrium, there is no heat current, and the energy density in terms of phonon number is

$$U = \sum_{q,b} n_{q,b}(\omega_{qb}) \hbar \omega_{q,b} / V \qquad (18.41)$$

The sum is over the wave vectors in the lowest Brillouin zone and over the acoustic and optical branches; V is the volume. The phonon number is the Bose–Einstein expression

$$n(\omega) = \frac{1}{e^{\hbar\omega/k_B T} - 1} \tag{18.42}$$

Here, k_B is Boltzmann's constant and T is the temperature.

18.3.1 Bulk Modes

Where there is a temperature gradient, there will be a flow of energy from where the energy density is higher to where it is lower. For a given lattice mode specified by wave vector and frequency, a higher energy density implies a higher phonon occupancy. Let the temperature gradient be along the x direction. A thermal current can be defined at $x = x_0$ in terms of the spatial difference of energy density and the velocity, v, associated with the mode:

$$j_x = v_x[U(x_0 - \ell\cos\theta) - U(x_0)] \tag{18.43}$$

Here $v_x = v\cos\theta$, and ℓ is the mean-free path. Assuming the mode velocity to be independent of direction and the mean-free path to be a constant, then, to lowest order, we can take the 'upstream' energy density to be

$$U(x_0 - \ell\cos\theta) = U(x_0) - \frac{dU}{dx}(\ell\cos\theta) \tag{18.44}$$

The thermal current is then

$$\begin{aligned} j_x &= -v\ell\frac{dU}{dx}\int_0^\pi \cos^2\theta\sin\theta\, d\theta\, 2\pi/4\pi \\ &= -\frac{1}{3}v\ell\frac{dU}{dT}\frac{dT}{dx} \end{aligned} \tag{18.45}$$

The thermal conductivity is

$$\kappa = \frac{1}{3}v\ell\frac{dU}{dT} \tag{18.46}$$

The total specific heat (dU/dT) involves a sum over all lattice modes. Optical modes, having a small velocity, may be excluded from contributing. The specific heat for a given acoustic mode is

$$U \approx \int_0^\infty \frac{\hbar\omega}{e^{\hbar\omega/k_B T} - 1}\frac{1}{V}q^2\,dq\,\frac{4\pi V}{(2\pi)^3} \tag{18.47}$$

Adopting the Debye model for acoustic modes, we put $\omega = vq$ and change the variable to $x = \frac{\hbar v}{k_B T} q$:

$$U = \frac{\hbar v}{2\pi^2} \left(\frac{k_B T}{\hbar v}\right)^4 \int_0^\infty \frac{x^3}{e^x - 1} dx \tag{18.48}$$

The integral can be evaluated noting that $\frac{1}{e^x - 1} = \sum_{n=1}^\infty e^{-nx}$ and $\sum_{n=1}^\infty \frac{1}{n^4} = \frac{\pi^4}{15}$, thus

$$U = \frac{\pi^2}{30} \frac{(k_B T)^4}{(\hbar v)^3} \tag{18.49}$$

The contribution to the thermal conductivity is

$$\kappa = \frac{2\pi^2}{45} v \ell k_B \left(\frac{k_B T}{\hbar v}\right)^3 \tag{18.50}$$

Contributions come from the three acoustic branches. In the isotropic model there is on LA and two identical TA branches. The total thermal conductivity is

$$\kappa = \frac{2\pi^2}{45} \ell k_B \left(\frac{k_B T}{\hbar}\right)^3 \left(\frac{1}{v_{LA}^2} + \frac{2}{v_{TA}^2}\right) \tag{18.51}$$

The assumption that the mean-free path is a constant can be valid only at very low temperatures, when umklapp anharmonic processes vanish, and momentum/energy relaxing collisions are determined by imperfections and boundary scattering. The conductivity then rises as T^3. Umklapp processes require at least one participant to have a sufficiently high energy/momentum, that is:

$$n(\omega) \approx \frac{1}{e^{\hbar \omega_Z / k_B T} - 1} \approx e^{-\hbar \omega_Z / k_B T} \tag{18.52}$$

The conductivity falls exponentially with increasing temperature. At high temperatures, $n(\omega) \to k_B T / \hbar \omega$ and $\ell = v\tau$, where τ is the relaxation time. The conductivity can be expected to fall following a weak power law.

This brief account encapsulates the main features of what is actually observed. Since it assumes the existence of bulk-like modes (which exist only in theoreticians desire for simplicity), it is pertinent to consider the role of normal acoustic modes.

18.3.2 Normal Modes

Let us consider the normal acoustic modes in an isotropic non-polar slab. Confinement is total, and we focus on the flow of heat in the x direction. The normal modes are the sTA, the pTA–LA hybrid (frequency of the TA mode), and the LA–pTA hybrid (frequency of the LA mode), as discussed in Chapter 6. Each mode has a pattern of displacement in the z direction that propagates along the x direction with a slab velocity, v, determined by its dispersion relation.

In the case of the s modes, the dispersion relation (in the Debye model) is

$$\omega^2 = v_T^2(q_x^2 + q_z^2)$$

$$q_z = \begin{cases} 2n\pi/L & \text{symmetric} \\ (2n-1)\pi/L & \text{antisymmetric} \end{cases} \tag{18.53}$$

The slab velocity is defined by $\omega = vq_x$:

$$v^2 = v_T^2 \frac{q_x^2 + q_z^2}{q_x^2} \tag{18.54}$$

A particular mode characterized by n and q_x makes a contribution to the heat current:

$$j(n.q_x) = -v(n, q_x)\ell(n, q_x)\frac{dU(n, q_x)}{dT}\frac{dT}{dx} \tag{18.55}$$

The energy density is

$$U(n, q_x) = \frac{\hbar v(n, q_x)q_x}{\exp[\hbar v(n, q_x)q_x/k_B T] - 1}\frac{1}{V} \tag{18.56}$$

At low temperatures the mean-free path can be taken to be constant, of order of the z dimension of the slab. For a given n, q_x can vary between the limits q_1 and q_2, where $q_1 = 0$ and $q_2^2 = q_{zb}^2 - n^2\pi^2\backslash L^2$, where q_{zb} is zone-boundary vector. The current associated with mode n is

$$j(n) = -\ell\frac{dT}{dx}\frac{d}{dT}\int_{q_1}^{q_2} \frac{\hbar v^2(n, q_x)q_x^2}{\exp[\hbar v(n, q_x)q_x/k_B T] - 1}\frac{1}{V}\frac{q_x dq_x d\phi\sigma}{(2\pi)^2} \tag{18.57}$$

A neater expression can be obtained with a change in variable:

$$x = \frac{\hbar v_T}{k_B T}\sqrt{q_x^2 + n^2\pi^2/L^2} \tag{18.58}$$

$$j(n) = -\frac{2\ell}{\pi L} \frac{k_B^4 T^3}{\hbar^3 v_T^4} \frac{dT}{dx} \int_{x_1}^{x_2} \frac{x^3}{e^x - 1} dx \qquad (18.59)$$

$$j = \sum_{n=1}^{q_{zb}L/\pi} j(n)$$

The limits are

$$x_1 = \frac{\hbar v_T}{k_B T} \frac{n\pi}{L}, \quad x_2 = \frac{\hbar v_T}{k_B T} q_{zb} \qquad (18.60)$$

The law of the cube of the temperature is recovered. Further progress must be numerical. However, as for lifetime, the effect of hybridization is small.

The current associated with the pTA/LA and LA/pTA hybrids is made less analytically accessible because of the more complicated dispersions (equations (6.7) and (6.10)). The calculation requires numerical methods from the outset. But there is no reason to suppose hybrids will give significantly different results from that obtained by bulk modes.

In the cases of quantum wells and quantum wires with non-rigid boundaries, the confinement of acoustic modes is less restrictive, suggesting smaller effects of hybridization.

References

Adachi S. (1985) J. Appl. Phys. **58** R1.
Akero H. and Ando T. (1989) Phys. Rev. **B40** 2914.
Anderson D.R., Babiker M., Bennett C.R., Zahklenuik N.A., and Ridley B.K. (2001) Phys. Rev. **B63** 24531: J. Phys. Condens. Matter **13** 5999.
Auld B.A. (1990) *Acoustic Fields and Waves in Solids*. Malabar, FLA: Krieger.
Babiker, M. (1986) J. Phys. C: Solid St. Phys. **19** 683.
Bannov N., Aristov V., Mittin V., and Stroscio M.A. (1995) Phys. Rev. **B51** 9930.
Banyai L. and Koch S.W. (1993) *Semiconductor Quantum Dots*. Singapore: World Scientific.
Beltzer A.I. (1988) *Acoustics of Solids*. Berlin: Springer-Verlag.
Bennett C.R., Ridley B.K., Zakhlenuik N.A., and Babiker M. (1999) Physica **B263–264** 469.
Bennett C.R., Tanatar B., Constantinou N.C., and Babiker M. (1994) Solid St. Comm. **92** 947.
Born M. and Huang K. (1954) *Dynamical Theory of Crystal Lattices*. Oxford: Clarendon Press.
Burt M.G. (1988) Semicond. Sci. Technol. **3** 739, 1224.
Burt M.G. (1992) J. Phys. Condens. Matter **4** 6651.
Cauchy A.L. (1822) (see e.g. A.E.H. Love, *A Treatise on the Mathematical Theory of Elasticity* 4th edn. 1927; New York: Dover 1944).
Chamberlain M.P., Cardona M., and Ridley B.K. (1993) Phys. Rev. **B48** 14356.
Comas F., Tralero-Giner C., and Canterero A. (1993) Phys. Rev, **B47** 760.
Constantinou N.C. and Ridley B.K. (1994) Phys. Rev. **B49** 17065.
Datta S. (1995) *Electronic Transport in Mesoscopic Systems*. Cambridge: Cambridge University Press.
Delves R.T. (1959) Proc. Phys. Soc. London **73** 572.
Dyson A. and Ridley B.K. (2012) J. Appl. Phys. **112** 063707.
Enderlein R. (1993) Phys. Rev. **B47** 2162.
Fang, F.F. and Howard, W.E. (1967) Phys. Rev. **163** 816.
Fedorov A.V. and Baranov A.V. (1996) Zh. Eksp. Teor. Fiz. **110** 1105.
Fletcher K. and Butcher P.N. (1972) J. Phys. C. **5** 212.
Fomin V.M., Gldilin V.N., Devreese J.T., Pakatilov E.P., Balaban S.N., and Klimin S. N. (1998) Phys. Rev. **B57** 2415.
Foreman B.A. (1995) Phys. Rev. **B52** 12260.
Foreman B.A. (1998) Phys. Rev. Lett. **80** 3823.
Foreman B.A. and Ridley B.K. (1999) Proc. ICPS24 CDROM Section V-E3.
Fuchs R. and Kliewer K.L. (1965) Phys. Rev. **140A** 2076.
Gold A. and Ghazali A. (1990) Phys. Rev. **B41** 7626.
Huang K. and Rhys A. (1950) Proc. R. Soc. **A424** 406.
Keating P.N. (1966) Phys. Rev. **145** 637.
Kim M.E., Das A., and Sentura S.D. (1978) Phys. Rev. **B18** 6890.
Klein M.C., Hache F., Ricard D., and Flytzanis C. (1990) Phys. Rev. **B42** 11123.
Klemens P.G. (1966) Phys. Rev. **148** 845.

Klimin S.N., Fomin V.M., and Bimberg D. (2008) Phys. Rev. **B77** 045307.
Knipp P.A. and Reineker T.L. (1992) Phys. Rev. **B45** 9091.
Krumhansl J.A. (1965) *Lattice Dynamics* (ed. R.F. Wallis, Oxford: Pergamon Press, 298).
Kunin I.A. (1982) *Elastic Media with Microstructure I*. Berlin: Springer-Verlag.
Lamb H. (1910) *Dynamical Theory of Sound*. London: Arnold.
Leburton J.-P. (1984) J. Appl. Phys. **56** 2850.
Lee J. and Spector H.N. (1983) J. Appl. Phys. **54** 6989.
Lee J. and Spector H.N. (1985) J. Appl. Phys. **57** 360.
Lee S.-C. and Galbraith I. (1999) Phys. Rev. **B59** 15796.
Love A.E.H. (1920) *Mathematical Theory of Elasticity*. Cambridge: Cambridge University Press.
Lowe D. and Barker J.R. (1985) J. Phys. C: Solid St. Phys. **18** 2507.
Marini J.C., Stebe B., and Karthauser E. (1994) Phys. Rev. **B50** 14302.
Martin R.M. (1970) Phys. Rev. **B1** 4005.
Merlin R., Güntherodt G, Humphreys R. Cardona M., Suryanarayanan R., and Holtzberg F. (1978) Phys. Rev, **B17** 4951.
Mitin V.V., Kochelap V.A., and Stroscio M.A. (1999) *Quantum Heterostructures: Microelectronics and Optoelectronics*. Cambridge: Cambridge University Press.
Mori N. and Ando T. (1989) Phys. Rev. **B40** 6175.
Morse R.W. (1948) J. Acoustical Soc. America **20** 833; (1950) **22** 219.
Morse P.M. and Feshbach H. (1953) *Methods of Theoretical Physics*. New York, NY: McGraw-Hill.
Mowbray D.J., Cardona M., and Ploog K. (1991) Phys. Rev. **B43** 1598.
Nash K.J. (1992) Phys. Rev. **B46** 7723.
Perez-Alvarez R., Garcia-Moliner F., Velasco V.R., and Trallero-Giner C.J. (1993) J. Phys. Condens. Matter **5** 5389.
Rayleigh, Lord (1877) *Theory of Sound*. Macmillan, Bristol.
Richter E. (1982) *Diplomarbeit Universitat Regensburg*.
Riddoch, F.A. and Ridley, B.K. (1983) J. Phys. C: Solid St. Phys. **16** 6971.
Ridley B.K. (1992) Proc. SPIE **1675** 492.
Ridley B.K. (1993) Phys. Rev. **B47** 4592.
Ridley B.K. (1996) J. Phys. Condens. Matter **8** L511.
Ridley B.K. (2009) *Electrons and Phonons in Semiconductor Multilayers*. Cambridge: Cambridge University Press.
Ridley B.K. (2013) *Quantum Processes in Semiconductors*, 5th edn. Oxford: Oxford University Press.
Ridley B.K., Al-Dossary O., Constantinou N.C., and Babiker M. (1994) Phys. Rev. **B50** 11701.
Roca E., Trallero-Giner C., and Cardona M. (1994) Phys. Rev. **B49** 13704.
Sezawa K. (1927) Bull. Earthquake Res. Inst. **3** 1.
Sood A.K., Menendez J., Cardona M. and Ploog K. (1985) Phys. Rev. Lett. **54** 2111, 2115.
Srivastava G.P. (1990) *The Physics of Phonons*. Bristol: Adam Hilger.
Stoneham A.M. (1975) *Theory of Defects in Solids*. Oxford: Oxford University Press.
Stonely R. (1924) Proc. Roy. Soc. **A106** 416.
Strauch D. and Dorner B.J. (1990) Phys. Condens. Matter **2** 1457.
Stroscio M.A. (1989) Phys. Rev. **B40** 6428.
Stroscio M.A. and Dutta M. (2001) *Phonons in Nanostructures*. Cambridge: Cambridge University Press.

Stroscio M.A., Kim K.W., Yu S., and Ballato A. (1994) J. Appl. Phys. **76** 4670.
Stroscio M.A., Sirenko Yu M., Yu S., and Kim K.W. (1996) J. Phys: Condens. Matter **8** 2143.
Trallero-Giner C., Garcia-Moliner F., Velasco V.R., and Cardona M. (1992) Phys. Rev. **B45** 11944.
Usher S. and Srivastava G.P. (1994) Phys. Rev. **B50** 14179.
Vallée F. and Bogani F. (1991) Phys. Rev. **B43** 12049.
Wang S.Q. and Mahan G.D. (1972) Phys. Rev. **B6** 4517.
Wang X.F. and Lei X.L. (1994) Phys. Rev. **B49** 4780.
Wendler L. and Grigoryan V.G. (1988) Surface Science **206** 203.
Yu S., Kim. K.W., Stroscio M.A., Iafrate G.J., and Balato A. (1994a) Phys. Rev. **B50** 1733; (1994b) Phys. Rev. **B51** 4695; (1996) J. Appl. Phys. **80** 2815.
Zakhleniuk N.A., Bennett C.R., Constantinou N.C., Ridley B.K., and Babiker M. (1996) Phys. Rev. **B54** 17838.
Ziman J.M. (1960) *Electrons and Phonons*. Cambridge: Cambridge University Press.
Zucker, J.E., Pincuk, A., Chemla, D.S., Gossard, A., and Wiegman, W. Phys. Rev. Lett. **53** 12280 (1984).

Index

3DP model 120

A
abrupt-step model 53
absorption rates 174
acoustic displacements
  in lattices 29–33
  in linear chain of atoms 8
acoustic equations
  inhomogeneous material 32
  and optical equations 29–33
  uncoupled 31
  see also equations of motion, non-polar acoustic modes; equations of motion, polar acoustic modes
acoustic hybrids, single heterostructure 75–80
acoustic modes (waves) 15–23
  boundary conditions 19–21, 46–7
  conditions for transmission 19–21
  continuum theory 15–16, 24
  free standing slabs 20
  guided 96
  inhomogeneous material 19–21, 32
  interface 80–2, 95–6
  isotropic crystals 19
  non-polar, equations of motion 16–17, 19–20
  non-polar slab 62–6
  polar, equations of motion 34
  quantization 21–3
  quantum wells 91–6
  velocities 17–19
acoustic strain 16–17
  quantum wires 106

acoustic stresses 16–17
  quantum wires 106–7
  in zinc blende 38
anharmonic interaction 177
anharmonicity 173
annihilation operators 21, 22
antisymmetric modes 64–5, 86

B
bare-phonon rate 174
barrier, single heterostructure 69–72, 79
barrier modes
  quantum wells 90–1
  scattering rate associated with 139–40
Bessel functions 99, 101, 127
  spherical 111, 113, 128
Bessel's equation 99
Bloch function 119, 125
Bose–Einstein expression 180–1
boundary conditions 46–54
  acoustic modes 19–21, 46–7
  electromagnetic 54
  linear chain of atoms 7
  non-polar slab 61–2
  optical modes 48–53, 119–20
  reduced
    quantum wells 88–90
    single heterostructure 74–5

C
colloids 110
connection rules
  acoustic 9–10, 36, 43, 46–7
  electric 11
  electrons 124
  in matrix form 51
  optical 10, 36, 43, 48, 50–1

continuum theory 15–16, 24
coupled-mode
  frequency 155–6
coupled modes 153–72
  beyond the long-wavelength approximation 156–61
    degenerate case 161
    non-degenerate case 161–2
  long-wavelength 153–6
  plasma with hybrid 170–1
  quasi-1D cylindrical structures 171–2
  screening in quasi-2D structures 163–9
    degenerate case 166–7
    long-wavelength solutions 168–9
    non-degenerate case 167–8
coupled plasmon–phonon modes 122–3
creation operators 21, 22
cylinder, scalar and vector fields 56–7
cylindrical coordinates, quantum wires 98–101

D
DC model 1–2, 120, 134
Debye model 179, 182
delta function 131
diamond lattice, microscopic theory 25–8
dielectric continuum model see DC model
dielectric function
  IF modes 41, 90
  intra-subband transitions 164
  LO modes 155
  plasma with hybrid coupled mode 170
  quantum 156–60
dilatational modes 108

dilatational stress 61
dispersion 33
  hybrid model 121–2
  non-polar modes, in single
    heterostructures 70–1
  polar modes
    in single
      heterostructures 70–1
    in zinc blende 40–1
    quantum well 85–6
    quantum wires 103–6
dispersion relation 6, 8,
  11, 42

E
effective charge 156
effective mass, electrons
  119–20, 124
elastic constant tensor 37
elastic constants
  acoustic 16, 20, 44
  optical 25, 32, 48, 49
elastic energy density 19
elastic isotropy 19
elastic stress tensor 37
electric displacement, in zinc
  blende 37–8, 43–4
electromagnetic boundary
  conditions 54
electron–hole interaction
  145, 148–51
electron–lattice
  coupling 145–8
electron–phonon interaction
  quantum dots 145–52
  in quantum dots
    electron–lattice
    coupling 145–8
    excitons 148–52
electron wave vector 128,
  131
electrons
  confinement 124–8
  and polar optical phonons
    see polar optical
    phonons and electrons
  scattering rate 128
  emission rates 173–5
energy normalization 23
  hybrid modes in polar
    material 103–5
  quantum wells 88, 89, 91
  single heterostructure 73–4
energy of vibration 25
envelope function 24, 124,
  125

equations of motion
  atoms in diamond
    lattice 27, 28
  linear chain of atoms 6,
    7, 9
  non-polar acoustic
    modes 16–17, 19–20
  optical modes 34
  piezoelectric effect on 44
  plasma oscillations 153–4
  polar acoustic modes 34
  polar optical modes 40
  vicinity of interface 19
excitons 148–52

F
Fang–Howard function 126,
  129
field-effect transistors 69
field factor 84
fluctuation-dissipation
  theorem 158
forces between ions
  non-polar component 11
  polar component 11
form factor (G)
  quantum well 137–40, 165
  quasi-1D cylindrical
    structures 172
  single heterostructure
    130, 132–3, 164

G
Gauss's equation 39, 40
glass, coloured 110
guided modes (waves) 64–6,
  94, 96

H
Hamiltonian 21, 22, 129–30,
  136, 146, 175–6
HD model 120
Helmholtz equation
  quantum dot 110–11
  scalar and vector fields
    55–6
Hooke's law 15
hot phonons
  lifetime 173–84
    effect if hybridization on
      175–80
    thermal conductivity
      180–4
  polar optical phonons and
    electrons 122–3

Huang–Rhys (HR)
  factor 147, 149–51
hybrid model 69–72, 120,
  131–2, 134
hybrid modes in polar
  material, quantum
  wires 103–6
hydrodynamic (HD) model
  120

I
IF-like modes 87–8
IF modes see interface (IF)
  modes
inhomogeneous material
  acoustic modes 19–21, 32
  optical modes 34–5
  polar modes 43
interaction energy 146
interface acoustic modes
  (waves) 80–2, 95–6
interface (IF) modes
  quantum wires 101–2
  single heterostructure 72
  in zinc blende 41–2
interface waves 42, 81, 94
isotropic crystals, acoustic
  modes 19
isotropy
  elastic 19
  optical modes 34
  polar modes 43

K
Klemens process 177–80

L
LA waves 18–19, 20, 91, 108
Lamb waves 63–4, 94
Landau damping 158, 161
lattice anharmonicity 173
lattice constants 20
lattice dispersion 24
lattice dynamics, microscopic
  see microscopic lattice
  dynamics
lattice potential 99, 100
LC model 120–1
  see also hybrid model
leaky waves 92, 94
Legendre function 128
Lindhard expression 157–8,
  158–9

Index **191**

linear chain of atoms,
  vibrational
  properties 6–11
linear-chain model 120
LO waves (modes) 39, 40, 41
  cylindrical coordinates
    for quantum wires
    98–100
  dispersion 70–1
  p-modes 53
  polar material 103–6
  quantum dot 111–12, 114
  scalar potential 98–9
  scattering potential 129
  velocities 34, 41
  see also optical modes
long-wavelength coupled
  modes 153–6
long-wavelength model 24
longitudinally polarized
  acoustic waves see LA
  waves
longitudinally polarized
  optical waves see LO
  waves
Love waves 62–3, 92–3
Lyddane–Sachs–Teller
  relation 40

M

mass approximation 36
Maxwell's equations 41–2
microscopic lattice
  dynamics 24, 25
  diamond 25–8
mode patterns 86–7

N

non-polar slab 61–8
  acoustic modes 62–6
  boundary conditions 61–2
  optical modes 67–8
number operators 22

O

optical displacements
  in lattices 29–33
  in linear chain of atoms 8
optical equations
  and acoustic
    equations 29–33
  homogeneous
    crystals 32–3
  uncoupled 31

optical modes (waves) 24–36
  boundary conditions
    48–53, 119–20
  decoupled acoustic
    and optical
    equations 29–33
  inhomogeneous
    system 34–5
  isotropy 34
  microscopic theory of
    diamond lattice 25–8
  non-polar slab 67–8
  polar 38–40
  velocities 33–4
  see also LO waves; TO
  waves
optical phonons 25, 42, 97
  see also polar optical
  phonons
optical strain 25

P

p-modes 53, 61–2, 63, 93–5
  double hybrid pattern 67
  see also pLA modes;
    pT modes; pTA
    modes
permittivity tensor 37
perturbing potential 156–7
phonon-like frequency 156
phonon wave vector 128, 131
phonons
  decay 177
  hot 122–3
  remote 72–3
piezoelectric coefficients 38
piezoelectric effects 11
piezoelectricity 43–5
pLA modes 63, 78–9, 91,
  93–4
plasma oscillations, equations
  of motion 153–4
plasmon-like frequency 156
polar elements 37–8
polar modes, quantum
  dot 111–12
polar modes in zinc
  blende 37–45
  effects of dispersion 40–1
  inhomogeneous
    material 43
  interface (IF) modes 41–2
  optical 38–40
  piezoelectricity 43–5
  polar elements 37–8
  velocities 42–3

polar optical modes
  hybrid model for single
    heterostructure
    69–72
  in zinc blende 38–40
polar optical phonons and
  electrons 119–23
  coupled modes 122–3
  dispersion 121–2
  hot phonons 122–3
  models 120–1, 123
polarization vectors 22
potential, nanostructures
  125–6
pressure vertical (PV)
  waves 91
promoting mode 176
propagation velocity 93
pT modes 61–2, 63
pTA modes 63, 66, 69, 76–8,
  93–4
  see also shear vertical (SV)
  waves
PV waves 91

Q

quantization
  acoustic modes 21–3
  simple harmonic oscillator
    21–2
quantum boxes 116
quantum discs 115–16
quantum dots 56, 110–16
  arrays 145
  electron–phonon
    interaction in 145–52
  Helmholtz equation
    110–11
  polar double hybrids
    114–15
  polar modes 111–12
  spherical coordinates
    110–14
  TO modes 113–14
  wave function 127–8
quantum well modes 20
  scattering rate associated
    with 135–9
quantum wells 83–96
  acoustic modes 91–5
  barrier modes 90–1
  energy normalization 88,
    89, 91
  interface acoustic
    waves 95–6
  lifetime of LO mode 177–8

quantum wells (*continued*)
  reduced boundary condition 88–90
  scattering rate in 135–41
  screening 164–5
  triple hybrid 83–8
  wave function 126
quantum wires 56, 97–109
  acoustic strain 106
  acoustic stresses 106–7
  cylindrical coordinates 98–101
  dispersion 103–6
  free surface 108–9
  hybrid modes in polar material 103–6
  interface modes 101–2
  scattering rate in 142–4
  screening 170–1
  wave function 126–7
quasi-continuum theory 47

R
Raman scattering 120, 156
Raman spectroscopy 74–5
Rayleigh waves equation 66, 81, 96, 109
reduced hybrid theory 89
reduced mass 3
  linear chain of atoms 8
reduced-mass density 39
remote phonons 72–3

S
s-modes 53, 61–3
  see also sTA modes
scalar fields 55–8
  cylinder 56–7
  Helmholtz equation 55–6
  sphere 57–8
scalar potentials 55–6
  quantum well 136
  quantum wire 97
  LO modes 98–9
  single heterostructure 129–30
scattering, remote-phonon 72
scattering potential, hybrid LO modes 129
scattering rate
  electrons 128
  quantum wells 135–41
  quantum wires 142–4
  single heterostructure 129–34

Schottky-effect devices 69
screening potential 157, 163–4
screening in quasi-2D structures 163–9
Sezawa waves 96
shear horizontal (SH) waves 91
shear moduli 33
shear vertical (SV) waves 91
simple harmonic oscillator, quantization 21–2
single heterostructure 69–82
  acoustic hybrids 75–80
  energy normalization 73–4
  hybrid model for polar optical modes 69–72
  interface acoustic modes 80–2
  reduced boundary condition 74–5
  remote phonons 72–3
  scattering rate in 129–34
  wave function for electrons 126
sphere, scalar and vector fields 57–8
spontaneous emission rates 174–5
sTA modes 61–3, 69, 75–6, 92–3
  see also shear horizontal (SH) waves
Stoke's shift 147
Stonely waves 95
strain 15–16
  acoustic 16–17
  quantum wires 106
  optical 25
stress 15–17
  acoustic 16–17
  quantum wires 106–7
  in zinc blende 38
surface modes 65–6
SV waves 91
symmetric modes 64–5, 86

T
TA waves 18–19, 20, 91, 108
  see also shear horizontal (SH) waves; shear vertical (SV) waves
TE modes *see* sTA modes
tensors 37

thermal conductivity, and hot phonon lifetime 180–4
TM modes *see* pTA modes
TO waves 39, 40, 41
  cylindrical coordinates for quantum wires 100–1
  dispersion 70–1
  quantum dot 113–14
  s- and p-modes 53
  velocities 34, 41
  see also optical modes
torsional modes 108
transversely polarized acoustic (TA) waves 18–19, 20, 91, 108
transversely polarized optical (TO) waves *see* TO waves
triple hybrid, quantum wells 83–8

V
valence-bond model (Keating's) 24, 25, 26
vector fields 55–8
  cylinder 56–7
  Helmholtz equation 55–6
  sphere 57–8
vector potentials 55–6
  quantum wires 97–8
velocities
  acoustic waves 17–19
  optical modes 33–4
  polar modes 42–3
vibrational properties, linear chain of atoms 6–11

W
wave function 120, 125, 135–6
  Fang–Howard form 126, 129
wave vectors, linear chain of atoms 7
wells
  single hetereostructure 69–72, 79
  see also quantum wells

Z
zinc blende, polar modes in *see* polar modes in zinc blende
zone-centre 84

The manufacturer's authorised representative in the EU for product safety is Oxford University Press España S.A. of el Parque Empresarial San Fernando de Henares, Avenida de Castilla, 2 – 28830 Madrid (www.oup.es/en or product.safety@oup.com). OUP España S.A. also acts as importer into Spain of products made by the manufacturer.

www.ingramcontent.com/pod-product-compliance
Ingram Content Group UK Ltd.
Pitfield, Milton Keynes, MK11 3LW, UK
UKHW022153230426
12049UKWH00003BA/83